Upper Darby Public Libraries

Sellers/Main
610-789-4440
76 S. State RD

Municipal Branch
610-734-7649
501 Bywood Ave.

Primos Branch
610-622-8091
409 Ashland Ave.

Online renewals:

www.udlibraries.org
my account

Connecting you to literacy,
entertainment, and life-long learning.

Intelligently Designed

Intelligently Designed

How Creationists Built the Campaign against Evolution

EDWARD CAUDILL

UNIVERSITY OF ILLINOIS PRESS

Urbana, Chicago, and Springfield

1 2 3 4 5 C P 5 4 3 2 1

∞ This book is printed on acid-free paper.

Library of Congress Cataloging-in-Publication Data
Caudill, Edward.
Intelligently designed : how creationists built the campaign
against evolution / Edward Caudill.
page   cm
Includes bibliographical references and index.
ISBN 978-0-252-03801-3 (hardback)
ISBN 978-0-252-07952-8 (pbk.)
ISBN 978-0-252-09530-6 (ebook)
1. Creationism—United States—History. 2. Evolution (Biology)—
Religious aspects—Christianity—History. 3. Intelligent design
(Teleology)—History. 4. United States—Church history—20th
century.
I. Title.
BS651.C38   2013
231.7′6520973—dc23   2013018103

To Robert, Daniel, and Gretchen Caudill

Contents

Acknowledgments

I am indebted to several individuals for assistance in this endeavor. Paul Ashdown and Dwight Teeter, colleagues and coconspirators on other projects, provided valuable suggestions for improving the manuscript. Dzmitry Yuron, Natalie Manayev, and Ioana Coman helped mine that seemingly endless vein of popular media that were relevant to this topic. Foremost, Robert Caudill, who suffered several years of listening to the triumphs and frustrations associated with the research and writing of this book. He accompanied me to the Creation Museum, discovered web sites that had evaded me, and spent some late nights in the library digging up articles from publications that had been given up for lost by myself and the library staff. I'll always remember the midnight phone call: "Guess what I found!" Thanks to all of you.

Intelligently Designed

Creationism's Political Genesis

After World War I, divisions that had been boiling below the surface of a modernizing, but still agrarian, nation erupted. For some, the war showed what could happen in an industrial world gone mad. For others, the currents of disruptive new knowledge confirmed that some people would be left behind in their shells of quaint tradition. A Tennessee town became the period's most garish exemplar of divergent worldviews.

This is a history of creationism not just as a science-religion issue, but as a political movement that skillfully engaged the press with a campaign grounded in American myths. Contemporary creationism's political origin is the Scopes trial, which became the template—politically, scientifically, theologically—for all subsequent evolution-religion clashes. The twenty-first-century struggle is not just with Darwinism, but against broader social change. In the same way, the fight in the 1920s was not solely with evolution, but with modernism. Such an implausible concept as a 6,000-year-old Earth has won wide acceptance in the United States because creationists, in the twentieth century, appealed to individual rights of freedom of expression and religion, the heroism of rebellion, the virtue of individualism, and the allure of the "frontier" whether geographic or scientific. Creationists, in the latter-day image of William Jennings Bryan, deftly fitted themselves to these traditions, made them part of numerous campaigns, and won over a large proportion of the populace. Creationists simplified the complex and made the issue more appealing to more people and more amenable to a political-conflict news story or sound bite.

Young-Earth creationism has flourished in America since the 1920s, with the press central to that movement even before the Scopes trial. In subsequent decades, new media provided creationists new opportunities to tell their side of the story to a growing audience. Radio, television, and now the internet have reported the creationist campaign and influenced and legitimized the creationist message. Creationists have found adherents from among 30 to 50 percent of the U.S. population over the last half century. At the Scopes trial, Bryan and his antagonist, Clarence Darrow, both offered arguments appealing to a wider public and in terms that made their arguments durable. They appealed to cultural values such as egalitarianism and respect for science and technology. This approach, in turn, cast the debate in terms that were familiar to the popular press and its audience. Bryan and Darrow were not talking just to a jury in Dayton, but to a nation. Their ideas resonated, and their tactics endured. Creationism found its way into the political mainstream. Adopting Bryan's rhetoric, young-Earth creationists won a substantial following and a place in national discourse by employing familiar themes with broad appeal. Because creationists understood mass media, they knew how to persuade not just a congregation, but a culture.

Scopes-era antimodernism has turned into a campaign against a single aspect of biology: evolution. The campaign has remained quite energetic in spite of creationists' consistent losses in courts. In fact, the creationism/ intelligent-design campaign succeeded. Proof of success was measurable in public opinion polls, in actions of state legislatures and local school boards, even in winning a presidential endorsement in 2005. In 2008 and 2012, several presidential-nomination candidates endorsed creationism. This is a history of how creationism won so many converts.

Creationism's Chroniclers

Books that venture into the evolution-creation conflict would fill several library shelves. The topic has been approached in numerous ways—as a science-religion contest, as religious history, as legal history, or as background to the contemporary education-religion controversy. First among histories of religion and conflict is Ronald L. Numbers's *The Creationists: From Scientific Creationism to Intelligent Design*, which explored creationism from the nineteenth century and publication of *On the Origin of Species* through a 2005 court trial in Dover, Pennsylvania. Numbers's comprehensive history of creationist theology covered the fight with science as well as with other religious denominations. As he enumerated the nuances of creationist theology, Numbers showed how creationists fit "science" to belief in a 6,000-year-old Earth. His history followed creationists as they stepped from the pulpit to the political podium and treated the creationists' politicking in an even-handed fashion, avoiding endorsement or condemnation of their political goals. "Like evolution," he wrote, "creation could readily be turned to social and political ends," which included abortion, feminism, sex education, and teaching evolution in schools.[1]

Chris Mooney was more partisan, as is evident in the title of his book, *The Republican War on Science*. Mooney argued just what the title said— that the Republican Party has been at odds with science for several decades. In particular, he cited the George W. Bush administration with regard to stem-cell research and global warming. Mooney included intelligent design/ creationism among the acts of intellectual fraud that Republicans employed in the pursuit of particular policies, including "teach the controversy."[2] One particular aspect of creationist politics was thoroughly explicated by Barbara Forrest and Paul R. Gross in *Creationism's Trojan Horse: The Wedge of Intelligent Design*. They exposed the beginnings of the "wedge" strategy in the early 1990s, the origin of the plan for eventually getting creationism taught in public schools, and the particulars of activities to put the so-called *Wedge Document* into action, up to and including the 2005 trial in Dover.[3]

Eugenie Scott's *Evolution vs. Creationism: An Introduction* was more focused on conflict than Numbers, but it provided an easily accessible history of the issue. Writing for a more general audience, she was particularly intent on showing why creationism/intelligent design is not science.[4] Like Scott, Michael Ruse, a professor of philosophy, wrote a history of conflict, but he took a more scholarly approach to the fight. In *The Evolution-Creation Struggle*, he dissected the arguments of creationists against evolution and disassembled their logic. Ruse found the fight a uniquely American one and concluded that the contest is a metaphysical battle that won't vanish anytime soon.[5]

There are a number of good histories on the Scopes trial. The most thorough was Edward Larson's *Summer for the Gods: The Scopes Trial and America's Continuing Debate over Science and Religion*. His detailed account of events and personalities leading up to the trial, and a day-by-day narrative of the trial itself, showed how the trial influenced subsequent legal battles and arguments over religion and science. Ray Ginger's *Six Days or Forever: Tennessee v. John Thomas Scopes*, originally published in 1958, is a dated but sound history of the trial. Ginger pointed out that both Bryan and Darrow appealed to a Jeffersonian tradition of the wisdom of the common man.[6]

Michael Lienesch's *In the Beginning: Fundamentalism, the Scopes Trial, and the Making of the Antievolution Movement* viewed the Scopes trial and the 1920s as part of a swell of popularity for fundamentalism. Lienesch interpreted *Scopes* as social movement that influenced creationist thought, as part of a continuum of events up to twenty-first-century controversies about teaching intelligent design as an alternative to evolutionary theory. The Scopes trial, he wrote, succeeded in framing fundamentalism as antievolutionism for subsequent generations.[7]

George M. Marsden provided a broader cultural history in *Fundamentalism and American Culture: The Shaping of Twentieth-Century Evangelism*, which traced fundamentalism from its nineteenth-century origins in evangelical movements. He, too, found the critical years for fundamentalists to be post–World War I and the 1920s, when Bryan put himself at the head of the movement. Marsden concluded that the Scopes trial was a catastrophe for fundamentalists because it narrowed fundamentalism to antievolutionism and permanently affixed, with the help of the press, the obscurantist label to fundamentalism.[8]

In a more contemporary vein, both Matthew Chapman and Laura Lebo wrote detailed journalistic accounts of the 2005 creationism-versus-evolution trial in Dover. Chapman, in *40 Days and 40 Nights*, directed a bit more ire at the creationists than Lebo, who was a resident of Dover and reporter for the local newspaper, the *York Daily Record*. Lebo's *The Devil in Dover: An Insider's Story of Dogma v. Darwin in Small-Town America* recounted the beating that creationists took in the trial, but her contempt was directed at the local creationists for their blatant attempts to skirt the law, for hypocrisy (one of the creationist defendants lied under oath about his actions), and for the divisiveness that the furor created in the community. Chapman also detailed events of the trial, and he found the intelligent-design case to be both bad science and bad theology.[9] Edward Humes, in *Monkey Girl: Evolution, Education, Religion, and the Battle for America's Soul*, considered the trial part of a culture war in America, with this particular aspect of the war going back to Scopes. Humes believed the media culpable in promoting intelligent design simply by covering the so-called controversy. He wrote, "Kitzmiller [the plaintiff in the Dover case] became at root everything the original Scopes trial started out to be but was not," especially since scientists were able to testify in 2005, something they were not able to do in 1925.[10]

Several histories followed earlier court cases concerning the teaching of creationism or intelligent design in public schools. Marcel C. LaFollette, in *Creationism, Science, and the Law: The Arkansas Case*, dealt with events leading up to and the arguments concerning the important 1981 Arkansas case, *McClean v. Arkansas*, in which the federal district court ruled that "creation science" was religion, not science. In *Trial and Error: The American Controversy over Creation and Evolution*, Edward Larson began with cases prior to *Scopes* and went up to *McClean*. Dorothy Nelkin's case history, *The Creation Controversy: Science or Scripture in the Schools*, was more broadly drawn, and also went up to the *McClean* case.[11] All of these case histories sorted through the legal nuances, the mazes of state and federal law, and the implications of the rulings. All three authors, obviously, wrote the works in the 1980s. Scott, Forrest and Green, Numbers, and the books on Dover are up to date on cases, but without the legal minutiae. The most current book on the legal issues is Frank Ravitch's *Marketing Intelligent Design: Law and the Creationist Agenda*, which included a legal history and plunged into the philosophical and scientific arguments surrounding intelligent design. Creationists, in Ravitch's words, marketed themselves to contemporary American culture. Though he found creationism and intelligent design neither scientific nor convincing, Ravitch warned of underestimating the movement at the risk of seeing scientific progress diminished under a new set of rules.[12]

Almost all of these books recognized, or at least alluded to, politics and various media in the history of creationism. None, however, dealt with creationism/intelligent design per se in the press, including the ways in which creationists appealed to the press and how creationists turned the attention into a movement. Numbers, Lienesch and others acknowledged the role of media in evangelicals' resurgence in the 1930s, via radio, and the 1950s, via television. Larson, in *Summer for the Gods*, detailed the press reporting and its impact on the event. Tona Hangen, in *Redeeming the Dial: Radio, Religion, and Popular Culture in America*, was one of the few authors who put both the media—radio in this case—and the audience in a broader cultural history. Hangen showed not only that radio arrested the decline of fundamentalism after the Scopes trial, but actually influenced the nature of twentieth-century fundamentalism.[13] Like Hangen, Mary Beth Swetnam Mathews tied the press to cultural outcomes for fundamentalists. In *Rethinking Zion: How the Print Media Placed Fundamentalism in the South*, she connected the negative images of the South in the press, especially those in the Northeast, to subsequent negative portrayals of religion in the South, particularly fundamentalism. Even before Scopes, according to Mathews, the press viewed the region as backward, a bias that the Scopes trial exacerbated.[14] Most histories of religion and media have been biographical or institutional. In either case, religion was subsumed under an individual life story or an institutional history. The consequences for the audience and the culture usually were left for another time.

These works, taken as a whole, justify using the Scopes trial as the beginning point for a historical understanding of the modern-day creation-science conflict. This history adds two elements in particular to the creationism story to help explain the movement's success: the use of enduring cultural myths and the dexterous employment of mass media. In the Scopes trial, Bryan provided fundamentalists a model of campaigning. He drew on broadly held cultural myths, such as egalitarianism and the frontier spirit, to win acceptance of his ideas. He understood that dramatic confrontation in the public sphere, which was the courtroom in this case, was irresistible to the press, which in turn communicated his "values," those mythic ideals, to the public.

There are no cultural histories of creationism as a politicized and mediated twentieth-century movement, which may explain in part why there are few analyses of how such a scientifically improbable idea as a 6,000-year-old Earth actually won such a large following in a scientifically and technologically advanced society. Such success was not just a matter of creating events that would attract public attention, which would be necessary but not sufficient. The attention needed to be gained in a fashion that would win a sympathetic response. In particular, this meant putting the Scopes trial in an American mythic tradition that made an event or issue more amenable to press coverage and which translated into wider audience appeal. The task here is to do more than describe the Scopes template for creationism. The goal is to identify the template's component parts, which were American myths and values, and show how these ideas were communicated to win a growing constituency for nearly a century.

Dayton, 1925

In July 1925, John Scopes went on trial for teaching evolution in a public school. A few businessmen and town leaders in Dayton thought a challenge to the state law against teaching evolution would get some attention for the economically struggling area. Those few town boosters, meeting in Robinson's Drug Store, talked the 24-year-old high-school teacher into accepting an offer by the American Civil Liberties Union to defend a test case. The trial turned into a press spectacle as newspapers, wire services, and even a radio station with a mobile transmitter descended on the town. Darrow and Bryan provided the star power. Darrow's notoriety rested on his high-profile defenses of labor and radical causes and his fight against the death penalty the previous year. In the 1924 Leopold-Loeb trial, two wealthy Chicago youths were charged with the thrill killing of a schoolmate. They were found guilty, but Darrow kept them off death row. Bryan was a prominent progressive politician, three-time presidential candidate, and leader of the growing fundamentalist movement that had found its demon in Darwin. The trial became agnostic versus fundamentalist, liberal lawyer versus conservative politician. Darrow and Bryan each sought to portray the other as a fool, with the dramatic high point occurring when Bryan took the stand as an expert witness to defend a literal reading of the Bible. Darrow relentlessly challenged Bryan on using the Bible as a book of science. How could the Earth be only a few thousand years old? How did Joshua make the sun stand still? How did a flood wipe out all life on Earth, except for what was on Noah's ark? Darrow called them "fool ideas." Bryan steadfastly defended faith, tradition, and the Bible.

Scopes's conviction was a foregone conclusion. Besides, the defense hoped to appeal the case to the Supreme Court of the United States. At the end of the twelve-day trial, Scopes was convicted and fined $100. The defense had agreed to let the judge, not the jury, fix the penalty. On appeal, the Tennessee Supreme Court reversed the judgment because the law required that the jury impose the penalty. So there was no case for the defense to appeal. The case was over, but the fight was just starting.

The trial politicized fundamentalism, which became synonymous with antievolutionism. Bryan and Darrow built their respective cases around themes that became permanent fixtures in the evolution-creation battle. Egalitarianism and individual rights were among the ideas that resonated well in the larger culture. Mixing religion and politics was nothing new in America, but the main reservations since Jefferson's time have been about mixing religion with government. Bryan represented evangelical republicanism, old-line Protestants who defined public virtue as submission to God's moral law as revealed in the Bible. That same evangelical theology exhorted believers to intervene in public life.[15]

The trial aggravated the tension between biblical literalists and modernists, especially the latter's critical and historical approaches to understanding the Bible. Objecting to "modernist criticism," though, would not engage a large constituency. When fundamentalists made evolution the nemesis of literalism, they turned an argument against an abstraction into a political fight clearly focused on public schools and children. Subsequently, creationists continued to win converts because the terms of the evolution debate remained very simple and essentially unchanged since the 1920s. Bryan's accomplishment was to fuse populist politics with evangelical and democratic traditions. Creating a political debate was important because it made the controversy more appealing to and relevant to a larger public. Furthermore, creating an event or spectacle meant a larger audience. Bryan not only made fundamentalism and antievolution synonymous, but he put them at core of what, by 2013, had become nearly a century-long campaign.

Bryan drew on ideals shaped by national myths of America as the apotheosis of a Christian republic, stirring in individualism, antielitism, and Jeffersonian egalitarianism. Darrow went to the same well of mythologies and also drew out individual rights. The modern-day creationist movement uses the same historical mythologies that inspired Bryan and Darrow, particularly Bryan's anti-intellectual impulse that enshrined the nobility of the common man and the righteousness of any decision or idea arrived at democratically. Creationists, in that Bryan tradition, adroitly have cast themselves as Jeffersonian egalitarians, antielitists and rebels forsaking convention and embracing new frontiers in science. At the same time, the antievolutionists stripped any scientific or theological nuance from the debate for the sake of a simple, two-sided battle. This made the controversy more amenable to public consumption. The press helped legitimize "creation-science" by putting it, in the name of fairness and balance, on equal footing with science.

In Bryan and Darrow, abstractions became real people in a real event as they scripted an ideal drama for dealing with such complexities in the court of public opinion: simplify, personalize, confront. In this respect, the trial's significance lay not in the resolution of any issues—at which it was an abysmal failure—but in the fact that it was a successful media event. It was both farcical—staged in self-anointed "Monkeyville"—and solemnly serious. The event was irresistible to the press and public. The same issue remains irresistible, erupting with regularity in state legislatures, local school boards, and even among presidential candidates.

The enduring influence of Scopes was evident in the 2005 creationism trial in Dover, which was a political story about one aspect of the culture war—a term born of 1980s politics—with one side being "conservative Christians" or "conservative Republicans." The labels were interchangeable. The other side was whoever opposed them. Media used a familiar frame for the science-religion fight in Dover: a two-party political contest rather than an issue with an infinite number of perspectives, from atheism to biblical literalists/antievolutionists. News values of fairness and balance fit naturally into the broader political-story template. The political story had only two sides, rather than innumerable ones. Even debate over the definition of "theory" took on a political tint by invoking free speech as part of the controversy. So the press got a story, made it comprehensible to the larger audience by simplifying it, and annoyed both sides in the process. Seen in this light, the success of the irrational in American culture becomes more understandable.

From Fundamentalist to Creationist

Creationism has a variety of meanings. It is used in this book in its Scopes-trial origins: belief in a young Earth that is 6,000 to 10,000 years old, and biblical literalism. Creationism also has been deemed "creation-science" and, more recently, "intelligent design." All of them inject supernatural causality into the study of science. "Intelligent design" may even accommodate an old Earth. Still, intelligent design is another term for creationism, as Forrest and Gross demonstrated in their history and critique of creationism and intelligent design.[16] Among the many contemporary fundamentalist ministries, most of them embrace intelligent design and belief in a young Earth. One of the largest twenty-first-century fundamentalist ministries in the United States is Answers in Genesis, which aggressively promotes young-Earth theology and biblical literalism in its publications, web sites, lecture tours, and its Creation Museum.

Answers in Genesis and other creation-science advocates differentiate between creationism and intelligent design. They deem "creationism" biblical literalism. "Intelligent design" is literalist by implication, and may or may not require belief in a young Earth, but ultimately relies on God as an explanation. The issue, more than a definition of creationism, is the definition of science. Creationists and intelligent-design proponents both have redefined science to include supernatural explanations, including science as it relates to public policy. That almost inevitably means the fight against teaching evolution in public schools.[17]

The Twenty-First-Century Campaign

Creation-science has not been taken seriously as science in more than a century. As theology, it is an old idea in new attire. It provides no knowledge or insight for theology or science. Creationists' twentieth-century innovation, though, was political—winning a very large constituency to a discredited idea. The contemporary debate is incomprehensible, if not ridiculous, when cast as a scientific issue. But if viewed as an extended partisan argument about public policy, the debate becomes comprehensible, especially in terms of creationism's success with great numbers of believers.

The 2005 events in Kansas and Dover illustrated both the appeal of intelligent design and the dexterity of its proponents. The movement has won national media coverage for a so-called scientific idea that has almost no support in the scientific community. In this respect, scientists' criticism of the press's coverage of the controversy was akin to criticizing networks because a political candidate looked good on camera. In spite of creationists' losses in both cases, the events revealed how creationism resonated in popular culture—it is a political issue with great media appeal. In Dover, creationists cast themselves as the rebels, fighting for their individual rights against the tyranny of an establishment mired in narrow orthodoxy. Creationists saw themselves in the American mythic tradition of strong-willed frontiersmen, trekking bravely into unknown territory, the establishment be damned. Reminiscent of the Scopes trial, a complex, contentious issue became a simple, two-sided fight in a Pennsylvania courtroom. The Dover case informed and entertained the larger national audience, which became the primary audience when the event slipped beyond control of local, elected officials who had attempted to inject young-Earth creationism into the school curriculum. Also reminiscent of Scopes, the 2005 cases had little discernible effect when it came to changing minds. Predictably, the press labeled the 2005 trial "Scopes 2," of which there must be hundreds by now if one includes the multitude of local science-religion flare-ups that never come to trial or national attention.

Creationism's political success stems from good organization, built around a well-defined, easily understandable issue and a committed constituency. Allies in the religious cause are manifestly political, such as those from the Christian Right that include Focus on the Family, Eagle Forum, Concerned Women for America, Coral Ridge Ministries, American Family Association, and the Alliance Defense Fund. Conservative Republicans are commonly in the group, too. Campaigning is a primal reflex of creationists, given their evangelical origins, and the tent revival has become a partisan rally, amplified by modern media. In contrast to creationists, scientists tend to be apolitical and narrow in interest in that conducting science is their profession, their concern. For them, engaging the public is a secondary concern.[18]

Recent presidential elections show creationism is politically viable. In 2008, two candidates touted their young-Earth creationist beliefs. During the Republican presidential primaries, Gov. Mike Huckabee of Arkansas minced no words on his position against evolution. Republican vice-presidential candidate Gov. Sarah Palin of Alaska offered antievolution views as part of her political credentials. Leading up to the 2012 presidential election, at least three major, if unsuccessful, contenders for the Republican nomination advocated antievolutionism: Palin, Texas Gov. Rick Perry, and Rep. Michele Bachman of Minnesota. Gallup polls since the early 1980s have shown anywhere from one-third to nearly one-half of Americans believe God created humans pretty much in the present form at one time within the last 10,000 years.[19] Thus, antievolutionism is a sound political strategy.

Myth and Message

Creationists have won press attention and converts, in part, by injecting their story with myths and values that hold broad cultural appeal. A myth may or may not be historically true, but as a story it teaches a lesson, clarifies issues, and explains problems, which often cannot be solved by reason or logic. A myth may resolve tensions and solve apparent contradictions in reality—i.e., how ancient fossils can exist in a 6,000-year-old Earth. Myths address anomalies in world views and simplify complex issues, as James Oliver Robinson has written.[20] Four myths that are particularly appropriate to the creationism/intelligent-design movement are the garden, the frontier, progress and science, individualism and egalitarianism.

The garden myth is a universal one, cutting across cultures and time. Because it is clearly biblical, the myth fits well—and necessarily—with the creationist campaign to make the symbolic garden a real place. In the nineteenth century, people believed the garden and creation were a real place and real event.[21] Darwin challenged that myth, and people were forced either to reject a literal Genesis or to reinterpret the creation story in light of new knowledge. Bryan cast the issue as science-modernism-city versus faith-traditionalism-agrarianism. He offered himself as a secular savior of his idealized republic of small farms and merchants, living the promises of Jefferson and Jesus. The creationist movement assumed the mantle of Bryanesque fundamentalism by trying to remain a sanctuary—a garden—from the threat of modernism, which evolution has represented for the movement from 1925 into the twenty-first century.

The creationist campaign has shrouded itself in apparent open-mindedness and adventure, claiming to venture intellectually to places—such as a 6,000-year-old Earth and seven-day creation—which they accuse mainstream scientists of avoiding. Their self-proclaimed adventurousness slips easily into the national frontier myth. In earlier times the reference point was the vast North American wilderness to be conquered, and out of which utopia would be carved. The frontier myth changed over time to represent opportunity, whether for wealth, democracy, or paradise. The frontiers to be conquered have ranged from geographic ones, including the West and outer space, to social ills (such as Lyndon Johnson's "Great Society"), to science and medicine (witness our multitude of "wars" on specific diseases, such as cancer and addiction).

Intelligent-design proponents seized on this mythic theme and cast themselves as the bold ones, the ones seeking to discover "new worlds," places that timid conventionalists—i.e., mainstream scientists—dare not go for fear of offending convention-minded colleagues. In such a scenario, the shunned outsiders turned into adventurers, moving into intellectual frontiers. This approach to the campaign meant creationists could move in concert with, rather than in defiance of, another American myth: science as progress. This mythology accelerated in the nineteenth century as belief in science merged with belief in progress.[22] As a result, an appeal to science and technology in American society was simultaneously an appeal to progress and forward-looking ideas. This created tension with fundamentalist Christianity. In the case of creationism, recognition of the status of science is seen in creationists' allusions to data and theories, particularly with the shift to "intelligent design" and the use of scientific language to demonstrate their progressivism.

Egalitarianism is inherent in the individualism that permeates American culture. Individualism involves strength of character, defying authority or establishment, being a pathmaker or a trailblazer. One can't be static and part of a bureaucratic structure in this particular myth. According to James Oliver Robinson, "A free American in pursuit of happiness . . . is *mobile*, is, has been, will be, in motion, *and* is defying authority, pathmaking (trailblazing), is in fact revolutionary. Moving, trailblazing revolution in defiance of establishment is the *way* of the proper American individual" [emphasis in original]. This appeal to individualism was illustrated in the actions, in 2005, of the Kansas state board of education, which rejected the "federalism" of mainstream scientists by altering the definition of science in order to accommodate intelligent design.[23]

In writing their own versions of history, creationists draw on these mythologies of the broader culture. They have redefined science in the spirit of Bryan to make it a political, rather than intellectual, endeavor. This rewriting of history and science thrives, and will continue to do so, because literalists have successfully created what Randall J. Stephens and Karl W. Giberson call a parallel culture in America. It is a culture with its own education system, from home schooling to colleges and universities; its own publishers; its own, near-closed social circles; and its own media system that includes print, web, and broadcast outlets for filtering information from the world outside.

Within this parallel culture, with its own media and information systems, the absurd becomes believable—i.e., Earth is about 6,000 years old, people and dinosaurs coexisted, a flood explains geologic strata, and so on—because creationists' history, religion, and science simply preclude the alternatives.[24]

Growing out of an evangelical tradition, creationists have familiarity with and skill in appealing to symbols, myths, and a powerful poetic tradition. Using symbols and myths is a daily practice. Historically, evangelicals have been quick to employ "new" media to their cause—radio in the 1920s and 1930s, television since the 1950s, and the internet in the twenty-first century. Such has not been the case for most professional scientists. Scientists understand science, and argue the issue from perspectives of validity and reliability, concepts that the public largely disregard or do not understand. Scientists are at a disadvantage in this political contest because their primary goal is to do science. Creationists evangelize, and science is of concern only inasmuch as it is a tool in promoting the primary message, which is religious. Scientists have continued to make a logical, empirical, scientific case against those who have no regard for logic, empiricism, or science.[25]

Creationists have drawn mainstream media to the issue by taking advantage of professional standards that obligated the press to cover court cases, to report on candidates and platforms, and to air public controversies at every level from local school boards to the U.S. Congress and presidents. Creationists simplified the issue and polarized the viewpoints, which made it easier for the press to report the controversy and for the public to consume it. The message invoked deeply held and deeply appealing cultural myths: America as the Garden, a nation of pathfinders and frontiersmen, a country of individualists. All the while, creationists embraced the national vision of science as progress.

A May 2012 Gallup poll showed that 46 percent of Americans believed God created human beings in their present form at one time within the last 10,000 years. The number has been fairly constant for the last three decades.[26] The appeal to cultural values and dexterity with media are part of the explanation for the phenomenal success of creationism in such a scientifically advanced society.

1

The Genesis of Young-Earth Creationism

[Scopes] is here because ignorance
and bigotry are rampant.
—Clarence Darrow, July 1925

Evolution is at war with religion.
—William Jennings Bryan, July 1925

Antievolutionism did not spontaneously generate itself in 1925, the
year of the Scopes trial, though one might think so in reading press accounts of subsequent innumerable cases involving teaching evolution in
public schools.[1] All of them seem to be "Scopes 2." There is justification
for the now clichéd label attached to cases involving religious objections to
teaching evolution in public schools. This is a result of the Scopes trial—the
first one—having become more than cultural shorthand for the evolution-
religion issue. The case also became a template for subsequent clashes over
the irreconcilable issue. That template includes

- Public schools as the battleground of choice.
- Winning media attention.
- Iconic figures, Clarence Darrow and William Jennings Bryan, whose
  arguments about the nature of science and appeals to cultural values
  and myths have endured.
- A political fight, as opposed to a philosophical, theological or scientific one. The issue usually erupts with elected local boards or state
  legislatures. Now, there are national organizations devoted to campaigning for young-Earth creationism, such as Answers in Genesis
  (AiG) in Cincinnati, Ohio, and the Center for Science and Culture, a
  branch of the Discovery Institute in Seattle, Washington.

The political dimension of creationism versus evolution is important be-cause scientists and the public so often have attempted to explain or under-stand these clashes as science-religion debates. Obviously, they are, in some respects. But the debate and the success of creationists remain incomprehen-sible to many because they ignore the political nature of antievolutionism. Like elections, campaigns against teaching evolution in public schools return with great regularity, the results of the last "election" cycle notwithstanding—i.e., that creationists eventually lost the legal case. But like any political cause or party they did not just go away. The popularity of young-Earth creationism is incredible, particularly considering that it has almost no credibility in the scientific community. "Success" is measured on the basis of national polls over the last several decades that have consistently showed anywhere from about one-third to one-half of Americans accepting a creationist-literalist reading of Genesis. However, this is not so much a failure of science as it is a triumph of politics. Though creationism is suspect science, it is a model of political activism, which took form at the Scopes trial.

Religious Objections

Those who have denounced Darwin have not always been theologically in-spired. During the late nineteenth and early twentieth centuries, some of the toughest critics of Darwin and *On the Origin of Species* (1859) and *Descent of Man* (1871) were scientists. Many scientists still saw science as a means of shedding light on God's workings in nature. Darwin set science on a new course by ignoring this traditional view of science's purpose and by discard-ing the presumed distinction between humanity and the rest of the animal kingdom.[2] In the twenty-first century, this remains an issue for many people.

A few decades after publication of *Origin*, a dedicated, conservative Chris-tianity coalesced around a defense of tradition, the threat of modernism, and a spirit of political reform. America's Social Gospel movement arose at a time of unregulated capitalism, abuses of labor, manipulations of markets, and the accumulation of huge fortunes by a few people. The Social Gospel was a Protestant attempt to sacralize Godless cities and factories.[3] The at-tempt to "save" the city was a restatement, perhaps even an affirmation in uniquely American terms, of the Old Testament Eden Myth. New ideas, new people, and new ways were slithering into the new-world garden. This new manifestation of the serpent, some believed, should be cast out of the garden in order to reclaim that sacred tradition. The emerging menace was complex and included modernists, who advocated new, critical approaches to art, literature, and history.

Modernism was an intellectual movement, imported from European universities, and thus was of little interest to most Americans. But any re-evaluation of sacred texts—the Bible, in this case—incensed many people. Evolution, another aspect of modernist thought, was fairly familiar to all, its implications easy to understand. Evolution's proponents said it applied to everyone, whether one rejected or accepted it. Two aspects of evolution repelled conservative Christians for several reasons. First, evolution was an implicit assault on a literal reading of Genesis. Second, it substituted a materialistic explanation of humanity's existence for a divine one. The objections were easy to understand, easy to communicate to others, and applied to the moment. Such immediacy and relevance to a large number of people meant evolution could be an issue with great political resonance.

The rapid growth of antievolutionism in the early twentieth century was more than a religious backlash to modern science. Since the early republic, evangelicals had seen themselves as having a mandate from God to transform society into a moral order. This idea ultimately collided with the social discord of late-nineteenth-century America, including labor strikes and the free-thought movement. Social order itself became an important part of a moral society. Universities were changing, less bound to religion and moving to empirical inquiry, particularly in the sciences.[4] Eventually, this view of the morality of order, which was part of a religious mission, was confronted with the implicit disorder or seemingly random variation that was part of Darwinism.

Organized antievolutionism arose from a larger stream of conservative theology in the late nineteenth and early twentieth centuries. A series of twelve pamphlets, *The Fundamentals*, published from 1910–1915 and distributed nationally, were defining documents for the movement. More than three million copies of each volume were distributed to pastors, professors, and theology students across the country. *The Fundamentals* launched the namesake religious movement with essays that addressed a variety of Christian doctrines, assailed modernist or "higher criticism" of the Bible, and promoted evangelism. The series contributed to fundamentalism's growing antievolutionism by devoting about 20 percent of its space to the subject, which ranged from more conciliatory, "liberal" views to uncompromising rejection of evolution. The lead essay of the seventh volume argued that evolution had been made materialistic by later Darwinists, and that Darwin himself did not exclude God's design as an explanation. The next generation, in other words, just went too far with the original idea. Essays in volume eight, however, rejected evolution, stated there was no universal law of development, and said transmutation of species was not possible. *The Fundamentals* espoused biblical inerrancy, which was not necessarily literalism. Evolution did not accommodate a literal reading of the Bible, and it was at odds with the fundamentalist vision of a planned and purposeful universe. At first, *The Fundamentals* did not provoke a lot of attention because they were not seen as particularly radical, though Christian fundamentalists later recognized them as the movement's beginning.[5]

Over the next decade, antievolutionists organized themselves, found leaders, and worked their way into the cultural mainstream with savvy media campaigns, appealing to religious tradition and national values. In a 1918 meeting in Philadelphia of what was later called the World Christian Fundamentalist Association (WCFA), fundamentalists saw in World War I the threat of modern industrialism realized, the culmination of amoral "survival of the fittest." In this maelstrom of ideas and events, fundamentalists became more militant, and any earlier tolerance for evolution began to disappear. Established in Philadelphia in 1919 and organized by William Bell Riley, an author of *The Fundamentals*, the WCFA was the movement's first formal organization.[6] In spite of the name, the organization's efforts were national.

The WCFA was not the only ingredient of a national political movement. *The Fundamentals* were, in effect, a policy platform published for mass consumption, and William Jennings Bryan was the party leadership. The pamphlets reached the grassroots level of the campaign by going out not just to church leaders but also to members. The publisher began offering the pamphlets for only 15 cents each, $1 for eight copies, and $10 for one hundred, encouraging their distribution in communities as well as churches. In the beginning of the third volume, the editors wrote that they found encouraging the receipt of more than 10,000 letters, from around the world, in support of the series.[7] At the same time, public education was growing, and many schools used textbooks that were favorable toward evolution. This meant evolution confronted an unprecedented number of people, usually as a matter of children being taught the idea in schools. It was a dual shock for many parents, whose children may have been the first generation to benefit from a higher level of public education. This accompanied a growing fundamentalist presence in politics, with thirty-seven states introducing antievolution legislation, which became law in Tennessee (1925), Mississippi (1926), Arkansas (1928), and Texas (1929). In 1922, the U.S. Senate went so far as to debate, but eventually reject, legislation to outlaw proevolution radio broadcasts.[8]

Religious conservatives found in evolution a subject that would drive a mass political movement, transforming the issue from a theological argument to a moral war in which the very soul of democracy was at stake. Public debates pitted notables from each side in dramatic, well-attended publicity events. The fundamentalists were good campaigners. For example, Bryan and other fundamentalist leaders knew the difference between Darwinism and social Darwinism, but did not differentiate. Such intellectual precision would have distracted from their message and been of little interest to an audience that generally was uninterested in such professorial fine points. So antievolutionists cast guilt upon the whole theory, and by extension upon materialism and modernism, for the decline of civilization. They also seized upon debates among scientists about the nature of evolution, whether it really was Darwin's natural selection or another mechanism, and cited the lack of perfect agreement among scientists as proof of the uncertainty about the whole idea of evolution. Any scientists expressing reservation about any aspect of evolutionary theory was further evidence.[9] This belied some ignorance about the very nature of science, especially the role of doubt and the necessity of a method that would provide a way to disprove a theory—i.e., if an idea cannot be subjected to empirical testing, it is not science. The indictment for scientific disagreement also revealed some hypocrisy in that fundamentalists ignored the fact that disagreement existed among theologians.

When Bryan emerged as the national leader of antievolutionism, he gave the movement its celebrity and its voice. He was widely known, and liked, and he knew how to lead a national movement—just as he knew how to conduct a national campaign. He had found an issue to put himself back in national politics, and antievolutionists had found a spokesman. His antievolution campaign took definitive shape in 1921, with a nationwide lecture tour that included "The Menace of Darwinism," in which he argued that evolution was un-Christian and unscientific. His evidence included the complexity of the eye, which he said was proof of God's design.[10] The popularity of the pamphlets, the growth of schools, and Bryan's injection of himself into the controversy made the issue appealing to the press because more people than ever were engaged in the topic, which Bryan simplified for easy consumption.

Reformers and Newsmakers

Bryan and Clarence Darrow are the historical personifications of the thesis and anti-thesis of the modern culture war, which explodes dramatically when it comes to teaching evolution in public schools. As the lead actors in the

Scopes trial, Bryan and Darrow defined the subsequent place of antievolutionism for fundamentalists. The Scopes trial was not just a reaction against Darwin and evolution, but against science in general. Leading the charge against teaching evolution in public schools was an opportunity for Bryan to combine his political experience, populist impulses, and pulpit principles. Though they became adversaries, both Bryan and Darrow were among the reform-minded Democrats of the late nineteenth century who looked for real change and acted on their own versions of profoundly moral principles to further the cause of humanity.

Like Bryan, Darrow identified with the common man, the underdog. Like Darrow, Bryan often associated with fringe political causes, such as socialism. Darrow first met Bryan at the 1896 Democratic National Convention. In *The Story of My Life*, Darrow praised Bryan's political ability and oration, especially the "Cross of Gold" speech, a response to Republican nominee William McKinley's gold-standard platform. Bryan electrified the convention with his delivery, eloquence, and words, including his famous closing: "You shall not press down upon the brow of labor this crown of thorns. You shall not crucify mankind upon a cross of gold!" Bryan went on to be nominated for president, with Darrow on down the ticket as a candidate for Congress. Darrow later wrote that he was relieved to lose. Bryan again won the nomination in 1900, and Darrow campaigned for him, motivated by Bryan's advocacy for Philippine independence. Bryan lost again. In 1908, their parting of ways started when Darrow refused Bryan's request for help in a third run for presidency. Darrow explained it as simply a matter of Bryan ignoring "dangerous problems." Bryan's alliance with a growing prohibition movement also irked Darrow.[11]

Bryan and Darrow both were experienced benders of public opinion well before the Scopes trial. Each was an accomplished politician and publicist. Darrow sought and nurtured publicity for himself and his causes from the earliest days of his legal career. He reveled in the headlines about his clients, which over the years included labor radicals, murderers, and socialists. His friendship with journalists was more than mere convenience or the sort of symbiosis so common in the news business between reporter and source. He had the journalistic impulse of the muckrake era to right wrongs, challenge authority, and defend the aggrieved. His relationship with two journalists in particular, H. L. Mencken of the *Baltimore Sun* and muckraker extraordinaire Lincoln Steffens, were philosophic kinships. Darrow and Mencken were fellow pessimists about the future of learning and enlightenment, and they held a mutual contempt for religion. At times, Darrow's writing even assumed a Menckenesque tone, as in "The Pessimistic versus the Optimistic View of Life": "What is a pessimist, anyway? It is a man or a woman who looks at life as life is. If you could, you might take your choice, perhaps, as to being a pessimist or a pipe dreamer. But you can't have it, because you look at the world according to the way you are made."[12]

Darrow and Steffens were fellow crusaders for labor and the common man, at least once in the same court case. They were longtime friends by the time of the trials of John and James McNamara for the bombing of the *Los Angeles Times* building in 1910. John McNamara was secretary-treasurer of the Structural Iron Workers Union, and James McNamara was a union sympathizer and activist. The bomb destroyed the building, killed twenty people, and wounded many others. Both Darrow and Steffens had doubts about the brothers' innocence, but they saw a larger truth at issue. When Steffens visited the brothers in jail—a privilege Darrow had granted him—Steffens told them he came not so much for the trial itself but to expose abuses of labor. The sentiment paralleled Darrow's. Steffens had a hand in engineering a plea bargain for the brothers, helping them avoid the death penalty. Darrow negotiated plea bargains, which were unusual in labor cases.[13]

A few years later, in July 1912, Steffens was the star witness for the defense in Darrow's bribery trial in which he was accused of bribing a juror in the McNamaras' trial. Asked if he were an anarchist, Steffens retorted was that he was "worse than an anarchist." He was a Christian: "It is more radical." In further testimony, Steffens's ideas about crime mirrored Darrow's: "I am trying to make a distinction between the crime that is merely done by an individual and the crime that is committed by an individual for a group which grows out of social conditions, and I think that those two lines of crime must be handled differently."[14] As for the McNamaras' trial, Steffens said Darrow pleaded the two brothers guilty because there was no other defense—not to avoid charges himself.[15]

Darrow's relationship with journalists reflected not just a shared philosophy but an appreciation for exposé as a tool for reform. His success as a newsmaker was among a plethora of traits that are common to successful public figures, but uncommon among scientists. He enjoyed the attention, shared values with his journalist friends, and was a talented communicator with a mass audience.

He used publicity in calculated fashion, fairly consistently adhering to principles of social justice, individual rights, and public education. His list of publications is impressive, particularly for a person who devoted many hours to his high-demand legal career. His books—authored, coauthored, edited—ranged from essay collections (*Persian Pearl: And Other Essays*, 1899), fiction (*Farmington*, 1904), social commentary (*Crime: Its Causes and Treatment*, 1922; *The Prohibition Mania*, 1927), to collections of agnostic writings (*Infidels and Heretics: An Agnostic Anthology*, 1929). He published in numerous periodicals, including *Current Topics*, the *Rubric*, *Liberal Review*, *American Mercury*, *Harper's*, *Vanity Fair*, *Scribner's*, the *Forum*, the *Nation*, the *Saturday Evening Post*, *Collier's*, *Rotarian Magazine*, and *Esquire*. He was eclectic, as he pondered labor issues, war in Europe, the League of Nations, capital punishment, mental illness, Zionism, divorce, prohibition, and the "absurdities of the Bible." He published articles, books and monographs, and pamphlets, and he took part in innumerable public debates and lectures over the course of about four decades from the 1890s to the 1930s.[16]

By comparison, Bryan was more than a friend to journalists—he *was* a journalist, owned a newspaper, and even held various editorial titles at different publications. Bryan had a newspaper and a great set of lungs. He used both in his exhortations as he attempted to harmonize the era's cacophony of political denominationalism in a chorus of civic Christianity. Bryan was a good fit for the fundamentalist cause not just for his conservative theology, but also for his prominence as a public figure. Far from being the bloviating rube of *Inherit the Wind* legend, he was eloquent and good with the press. Later in his career, he even identified himself as a journalist, which had some justification, given his newspaper, the *Commoner*, which was his foremost journalistic—or public relations—accomplishment. Its purpose was to promote Bryan, running his major speeches and reprinting articles he wrote for other publications. He edited the paper, aided by his brother Charles and Richard Metcalf, Washington correspondent for the newspaper and Bryan's ghostwriter. The *Commoner* was a campaign instrument with evangelical overtones. Bryan started hiring staff and arranging printing only a few days after his 1900 election defeat. In its first issue in January 1901, the newspaper called for a "revival" of the reform movement against monopolies and warned people not to fall into the same "backslidings" as the ancient Israelites by turning to the "doctrine of empires." It often read like a sermon on political issues. Bryan melded neatly his evangelical Protestantism and political Jeffersonianism with his appeal to commoners and scorn of the elite. By 1906 it circulated 145,528, according to *Rowell's American Newspaper Directory*, and was up to sixteen pages, half of that advertising. Bryan invited other publications to reprint any part of the *Commoner* for free. And they did, including major East Coast newspapers, helping him gain a larger audience than its already substantial circulation reflected.[17] By 1923, more than 100 newspapers carried his syndicated columns. In all, he was estimated to have an audience of 20 million to 25 million people annually.[18]

Bryan was editor-in-chief of the *Omaha World-Herald* in the summer of 1894, but it seemed largely a ceremonial post because the editorials attributed to Bryan were written by Metcalf. Publisher Gilbert Hitchcock carried out the editor's duties, as he undoubtedly saw good business in the alignment with the Democratic Party's rising star. Though editor only in title for the most part, Bryan eventually had some legitimate claims to being part of the profession. In 1912, he was a *New York World* reporter at the Republican convention in Chicago, and at the Democratic convention in Baltimore. It did not seem to bother him or the *World* that he also was a candidate.[19]

In addition to his printed pulpit, Bryan was a very popular public speaker throughout his career. His reputation as an orator was well established before his 1896 "Cross of Gold" speech. By 1895 he was earning $100 per speech. Not only was he talented, he was prolific, delivering more than 6,000 speeches over a thirty-year span in the public spotlight. His venues ran the gamut of campaign halls, revivals, civic events, high-school and college commencements, city parks, and trains. He was among the biggest names and best draws on the Chautauqua circuit from 1900–1920, when he made as much as $2,000 weekly from his lectures. Chautauqua and Bryan were good matches because the audiences tended to be small-town and rural, i.e., his political constituency, drawn to his agrarianism, antielitism, and antiurbanism.[20]

By the time of his post–World War I antievolution campaign, he had behind him several decades on the Chautauqua circuit. The crowds in the thousands were testament to his popularity. Once, he reportedly spoke unamplified to 100,000, who could hear him up to three blocks away.[21] In 1922, his lecture series at the Presbyterian Seminary in Richmond, Virginia, were a resounding success. The series, titled *In His Image*, was repackaged in book form and sold more than 100,000 copies.[22] He had a politician's instinct for maximizing publicity and a journalist's acumen for engaging an audience. Bryan's facility for winning press attention has been well documented, but this aspect of his contribution to the antievolutionism has been underappreciated.

From Quaint Religion to Fiery Politics

The public picture of fundamentalism changed rapidly over only a few years just prior to the 1920s and up to the Scopes trial. Fundamentalists moved from being seen as something of a theological oddity, conservative but not threatening, to being perceived as fanatics who wanted American culture to revert to the previous century. Well before the Scopes trial, the press was an important vehicle for the new image, with Mencken at the front: "They march with the Klan, with the Christian Endeavor Society, with the Junior Order of United American Mechanics, with the Epworth League, with all the rococo bands that poor and unhappy folks organize to bring some light of purpose to their lives."[23] Linking fundamentalists to the South and the Klan made the danger more palpable, moving them from oddball proselytes to menacing radicals.

The press had not always seen fundamentalists as backward fanatics. In 1910, at the meeting of the WCFA in Philadelphia, the *Inquirer* reported the group's opposition to "survival of the fittest" because it was contrary to Jesus's ideals. But the coverage was not alarmist and generally was balanced. As late as 1923, *Time* magazine reported dispassionately on an upcoming meeting of the General Assembly of the Presbyterian Church in Indianapolis and the argument between fundamentalists and "Liberals," which would include those modernists who looked critically at the Bible as history and literature. One side, *Time* said, claimed ideas such as evolution and the reign of law were revelations that could be "made friendly to the truths of Christianity," and the other side held these things "to be irreconcilable with Christianity as taught in the Bible."[24] The liberal *Nation* reflected a similarly even-handed view of that 1923 assembly, pointing out that Presbyterian modernists had gotten ahead of their denominations in an effort to keep their theology contemporary. The *Nation* found fundamentalism in "the main stream of Christian tradition while Modernism represents a religious revolution as far-reaching as the Protestant Reformation." In fact, it was the modernists, not the fundamentalists, who were outside the mainstream, the *New Republic* editorialized in January 1924. The magazine said fundamentalists were correct in believing that people who did not adhere to a denomination's beliefs simply did not belong to that denomination. Fundamentalists, according to the *New Republic*, were closer to the Protestant evangelical tradition that most Americans embraced.[25] When Harvard President Charles W. Eliot wrote on denominationalism for *Atlantic Monthly* in March 1924, he referred not to "literalists," but to those who espoused the "inerrancy" of the Bible in matters of faith. "Literalists" implied a thoughtless embrace of words, while "inerrancy" suggested something more theologically profound in consideration of the Word. Eliot found fundamentalists understood the literary aspects of the Bible, its use of poetic image, allegory, and metaphor.[26]

Bryan's brand of fundamentalism was somewhere between Dwight L. Moody's pragmatic, middle-class appeal and Billy Sunday's message of simple, traditional virtues from the Gospel. Moody stressed the love of God in his sermons, but condemned the "4 great temptations" that threatened society: theater, disregard of Sabbath, Sunday newspapers, and atheistic teaching. The last one included evolution. Moody, who died in 1899, influenced the whole subsequent generation of fundamentalists, including Bryan. Moody believed in biblical infallibility, but he cultivated "liberals," even though he disapproved of their ideas. Neither Moody nor Bryan concerned themselves with fine points of theology or doctrinal differences across denominations, but focused on Christianity as a progressive force in bettering society. The tolerance toward doctrine and focus on results also were characteristic of the more theatrical Billy Sunday. More staid conservatives recoiled from his stage extravagance, but his vaudevillelike techniques drew big crowds to revivals.[27] Viewed in this larger context, Bryan was not alone in reaching out to mass audiences, nor in his tolerance of doctrinal difference. Bryan was unique in the political accomplishment and skill he brought to his evangelizing.

Negative depictions began to seep into reporting as fundamentalism became increasingly associated with the South, more strident in tone and increasingly intolerant. The editor of *Century* magazine wrote in May 1923 that if Jesus were to return he would fear for His safety if He "were to fall into the hands of Fundamentalists." One Unitarian minister labeled fundamentalists "the religious Ku Klux Klan."[28] The accusations of violence and ignorance had already become more common as the summer of 1925 approached. The *Nation* called John Roach Straton, fundamentalist pastor at New York City's Calvary Baptist Church, a person so bent on revenge that he dared God to destroy New York City. A satiric article, by *New York Tribune* literary critic Heywood Broun, would have Straton call down God's vengeance and "smite" New Yorkers for watching baseball on Sunday.[29] Philosopher John Dewey equated fundamentalism with ignorance when he wrote for the *New Republic* in 1924 that fundamentalists "proclaim the infallibility of men who lived many centuries ago in periods of widespread ignorance, of unscientific methods of inquiry, of intolerance and persecuting animosity."[30]

The South was Bryan's political territory—more rural, traditional, religiously conservative, more threatened by inevitable cultural changes. By contrast, Darrow called Chicago home, where he socialized with the intelligentsia, pondered the great issues of the day, and indulged his conscience in just causes, especially labor and death-penalty cases. When the showdown between the kingdoms represented by Bryan and Darrow finally arrived, a Tennessee town with imported defenders of the wayward teacher seemed a made-to-order forum for any high-school drama production. Regional differences highlighted in the trial have always been a fact of American life. By the 1920s, popular perception, the press and radio, and radical racism exaggerated the differences and made them even more susceptible to dramatic exploitation.

An extraordinary literary movement also amplified regional differences as it helped ignite a new vitality for the idea of a unique Southern identity. Centered in Vanderbilt University in Nashville, "The Fugitives" were a literary enclave that redefined Southern identity and character. The small group included such luminaries as John Crowe Ransom, Allen Tate, Andrew Lytle, Robert Penn Warren, and Donald Davidson. Their essays, novels, poetry, and history defended Southern heritage and traditions. Ten years after their loose, initial organization, the Scopes trial reenergized their efforts. Five years after the trial, a broader group, which called itself the Agrarians, published *I'll Take My Stand*, a collection of essays in defense of Southern values. *I'll Take My Stand* was a reaction to the insults hurled at the South during and following the trial. In the book's Introduction, Ransom criticized "the cult of science" for its pretense and its blindness to nature's mysteries. The work defined Southern culture by its allegiance to agrarian values and its opposition to modernism.[31] It gave Southerners a positive and moral mythology, which complemented Bryan's Jeffersonian-agrarian vision. True to historical patterns, it seems, a critical aspect of the Southern identity was its formulation in opposition to something, which was the Northern industrial, urban landscape. An oppositional identity remains at the core of the contemporary culture war.

There was a dark side to the Southern identity, one that national media seized upon. A resurgent Ku Klux Klan and racial violence exacerbated the perception of the South as being apart from the rest of America. The *New York Times* reported in 1919 a series of church burnings in Georgia. The *Nation* lambasted Georgia for lynchings, and Texas for the killing of a 16-year-old boy, and it ran a collection of Southern newspaper headlines concerning a public lynching in Mississippi. Another popular magazine of opinion, the *Outlook*, reported that black people were being tortured and burned at the stake. In 1921, the *New Republic* reported the Florida town of Ocoee was the site of more than fifty killings of blacks because they voted in elections. Mobs burned the town of Rosewood, Florida, and lynched its residents, according to the *New York Times* and the *Washington Post*.[32]

By the mid-1920s, the American public was primed for a spectacle of ignorance from the Tennessee backwoods. The newspapers and magazines called the approaching trial "deliberate medievalism" (the *Independent*), a return to the middle ages (*Nation, Science*), an "Inquisition" (*Collier's, Christian Century*), "pre-Reformation" (*Outlook*), anticivilization (*New Republic*), and so on.[33] The public read of a region steeped in violence and populated with uneducated, gun-toting mobs. The image did not separate the area's surging fundamentalist movement from the violence. Just as fundamentalists had seized upon evolution to encapsulate their opposition to modernism, the "outside" world fitted antievolutionist sentiments to a caricature of an uncivilized, fanatically religious American subculture.

The press came to the Scopes trial to affirm perceptions of a subculture wallowing in ignorance and enamored of violence. Fundamentalists were a national target for a press primed to see the consequences of primitive religion. Thus, fundamentalists became a lightning rod for the ills of Southern culture. Already, the antievolution campaigns had succeeded, perhaps too well, as people outside the South came to see antimodernism/antievolutionism as the region's reigning mindset.

A few decades later, *Inherit the Wind* confirmed the legend. The 1955 play and 1960 movie recreated Bryan as a fanatical, babbling fool who finally collapsed in the courtroom. Many histories of the trial cast him as the loser, unable to answer Darrow's pointed questions concerning the illogic of biblical literalism. The movie's people of Dayton, Tennessee, were a mindless, marching mob—in diametric opposition to the historical reality of a town looking to provoke some promotional publicity for itself. In the play and movie, the Bryan and Darrow characters were cast in stark bad-guy and good-guy roles. Drummond (Darrow) became a crusader for civil liberties, even lambasting Hornbeck (the Mencken character) for ridiculing Brady's (Bryan's) religion. Both the play and the movie were hits. Numerous critics cited the gross misrepresentation of historical facts, such as softening Darrow's antagonism to religion, which often veered to outright hostility. The drama simplified Bryan's acceptance of "day-age" creationism, which was the idea that biblical days could have been long ages, and did not bother with the fact that social Darwinism was his real focus.[34] *Inherit the Wind* helped accelerate the politicization of creationism by starkly simplifying the issue, affirming and exaggerating bad-guy/good-guy figures in Brady and Drummond, and packing it neatly in a 128-minute cinematic morality tale. The real complexity of the fundamentalism-evolution conflict would have been poor fuel for political fires. This mythic version of the Scopes trial is far neater, far more entertaining than the historical reality of contrived publicity via an event with a foregone conclusion. The trial's dramatic high point—Darrow's grilling of Bryan—was something of a disappointment in reality because so many journalists already had left town, including

Mencken, and so were not there to publicize it. *Inherit the Wind*'s fictionalized version solved that, and had Bryan/Brady physically crumbling in defeat on the courtroom floor. With its leader defeated, fundamentalism was defeated. Great drama. Lousy history.

The Real Audience

The depiction of Bryan in *Inherit the Wind* was misleading for several reasons. First, Bryan was not speaking to the "elites" but to his constituents, the "common people." In the final day of the trial, he declared, "I am simply trying to protect the word of God against the greatest atheist or agnostic in the United States. (Prolonged applause.) I want the papers to know I am not afraid to get on the stand in front of him [Darrow] and let him do his worst. I want the world to know."[35] Second, he was not testifying so much as he was performing, giving a stump speech, which happened to be in a courtroom. Third, the loss was just a legal one, not a defeat in terms of defining and affirming what came to be an enduring national issue.

The issue still exists pretty much as Bryan framed it. Themes that persist to this day emerged from Bryan's remarks and testimony in the trial, and from his posthumously published *Last Message*, which would have been his closing argument. Several of those themes have become guiding principles of the contemporary creationist/intelligent-design movement, including the disdain of experts, which is tied to populist appeal. This theme branches in many directions, including anti-intellectualism, antielitism, and the Garden Myth that exalts country over city. In the courtroom, Bryan persisted in his argument that evolution was not science, but mere "guesses," which he equated with hypotheses. Such misunderstanding has persisted in the general populace, even to the point of pulling the word "theory" into the legion of synonyms for "guess" or "speculation" in the nonscience community. In the *Last Message*, Bryan's appeal moved to individual rights and the problem in a democratic society of beliefs being imposed upon one group by another. It is an effective argument, showing Bryan's political acumen. It is an American classic, too, going back to James Madison and Alexander Hamilton in their *Federalist Papers*, in which no. 10 raised the issue that the majority must not be able to tyrannize the minority. In this case, though, Bryan seems to have believed it was an elitist minority tyrannizing the majority.

He appealed to people anxious about the preservation of their traditions and place in American culture. Scientists, professors, the literati did not concern him. Bryan knew they would not align for him, no matter what. His

Last Message was aimed at the masses and defended their faith: "God may be a matter of indifference to the evolutionists and a life beyond may have no charm for them, but the *mass of mankind* [emphasis added] will continue to worship their Creator."[36] The address is as noteworthy for what it does not do as it is for the insight it provides to Bryan's thinking. Bryan largely ignored legal issues and said little of laws or the actions of John Scopes. That is because it was a campaign speech, aimed at an audience far wider than the jury box or the courtroom. Bryan again was the Chautauqua-circuit orator, the populist-newspaper editor, the campaigner. He was indifferent to scientific detail or philosophic nuance because to cite such things could have cast doubt on his own authenticity as an everyman. The widely circulated 15,000-word argument, which Bryan revised just after the trial to deliver as a speech across the country, summed up his considered arguments against evolution, as opposed to the exchange with Darrow during the trial. In the published message, Bryan said he was glad to have the trial in Dayton, before the "yeomanry." For him, the rural town was the superior venue. He dismissed the "insolent minority" that would interfere with majority rights, and the "oligarchy of self-styled 'intellectuals.'" There was, he declared, no need for expert testimony because any 14-year-old would be able to explain evolution. He insisted that the issue was "teaching guesses that encourage godlessness." He pointed out that Tennesseans valued science, but evolution was nothing more than "millions of guesses strung together." In reading from *Descent of Man*, Bryan planned to point out words "implying uncertainty" about the "brute hypothesis." He offered Darwin himself as evidence of the fact that evolution erodes faith. Bryan claimed that more than half of all scientists doubt or deny God. Evolution was an assault on Christianity and democracy. Darwinism eroded the moral social order, subordinating individual responsibilities to the dictates of genes and environment.[37]

Darrow's arguments also have persisted. One of his most common criticisms of Bryan and his antievolution constituency was that they were antieducation. But at times Darrow extended the criticism beyond fundamentalism and conservative Christianity to all of religion: "[H]e did not represent a real case; he represented religion, and in this he was the idol of all Morondom." Darrow put religion in opposition to learning, particularly science. He called the Scopes trial "this great waterloo of science."[38] Here, his tone anticipated later twentieth- and twenty-first-century antireligionists, such as Richard Dawkins, who found fault not just with literalism, but any and all religion, which Dawkins denounced as silly, ridiculous, even harmful.

Much of the mythology of the trial arose from the famous exchange on the witness stand. Darrow may well have triumphed that day. But the reality is more complex. Bryan and Darrow spoke to different audiences. Scientists and historians have tended to be in Darrow's audience, viewing the exchange on grounds of logical, empirical triumph. *Inherit the Wind* confirmed their verdict. Reading the trial transcript, it is easy to come to such a conclusion. Bryan failed to explain logically any number of miracles that Darrow threw at him, such as the presumed effect of Joshua making the sun stand still if Earth were rotating, or whether it was a whale or fish that swallowed Jonah. Bryan even admitted at one point that some of the Bible is not to be taken literally. But what may have been equally significant in the exchange is not that Darrow makes literalism illogical, but that Bryan stood his ground and defended the Word. Reading the transcript not as an evolutionist but as a defender of the faith, one could see Bryan confronting the threat, taking on science. He became the righteous rebel. This second audience may have seen faith persevere, as the empirical evidence was irrelevant to faith's defender. The audience to which Bryan spoke heard him defend the Bible and its adherents. Faith and the Great Commoner took a stand. In that Lost-Cause tradition, loss ennobled them.

In the Beginning, a Trial

Scientific Monthly in May 1923 said Bryan's popularity was in decline, but he remained popular in most of the South.[39] His appeal in the South was a matter of fundamentalist religion and conservative politics. Bryan and his ideas appear to have become even more popular with time. Just as the South was the center of the conservative religious movement in the early twentieth century, it was the center of a resurgent conservatism in the late 1970s and early 1980s with the emergence of the "Moral Majority." A culture war often split along regional lines with Southern states reliably Republican and conservative.

Bryan politicized a brand of fundamentalism steeped in his own interpretations of Christ and Jefferson. Bryan's underlying principle that truth, even scientific, was subject to the will of the people probably would not have set well with Jefferson. Darrow's progressivism was more agnostic about tradition, not rejecting it but simply disregarding it in the face of modern social issues and injustices as he embraced new approaches to understanding. The Scopes trial condensed their ideas about science and religion before a national audience, which each man sought, with enduring consequences.

2

The Contrarian and the Commoner

Darrow and Bryan

No scientific fact . . . can disturb religion, because facts
are not in conflict with each other.
—William Jennings Bryan

The human being who has the least intelligence
has the most faith.
—Clarence Darrow

Bryan and Darrow inhabited one of those peculiar moments in history when two individuals really did personify two major, antagonistic streams of thought in American society. Both had made their reputations in progressive causes, allied with the Democratic Party. But they saw different avenues to reform. Bryan believed religion informed all walks of life, including politics and science.[1] Darrow saw education, including science, and modernist thought as the way of changing an unjust society and improving the lot of the underclass.[2] Religion, he believed, too often was an impediment to enlightenment and just as corrupt as a capitalistic system that treated workers like little more than stockyard commodities. Both were at odds with the Gilded Age aristocracy. They sincerely sought to improve the lot of laborers, farmers, and others on the lower ends of the economy. Bryan had unshakable faith in the common man. Darrow called himself a pessimist, apparently with little faith in humanity, but he fiercely defended the rights of the common man. The two men's differences grew more pronounced with time.

History has been cruel to Bryan, the "boy orator" from Nebraska, three-time presidential candidate, and leading progressive. But he may deserve a scathing assessment. Rushing to the Scopes fray, he appears to have been motivated partly by principle, partly by vanity as he presumed to take on the greatest trial lawyer of the day. This from a man who had not practiced law in three decades. *Inherit the Wind* and journalist H. L. Mencken's anti-eulogy are the essentials of a history that often has diminished Bryan to a patron of ignoramuses. But he was a substantial figure because he helped define the Democratic Party for the era, to the point of writing the party platform five consecutive times from 1896 to 1912. In addition to his presidential runs in 1896, 1900 and 1908, he served as secretary of state under Woodrow Wilson. Throughout his political career, Bryan pitted himself against the intellectual elite, and the grinding, dirty factory of urban America. Biographer Michael Kazin stated, "Only Theodore Roosevelt and Woodrow Wilson had a greater impact on politics and political culture during the era of reform that began in the mid-1890s and lasted until the early 1920s."[3] The *Boston Daily Globe*, in Bryan's obituary, chronicled his national campaigns and offices, but set those against his final days, chugging about the Tennessee countryside, exhorting small-town crowds from a train.[4] Similarly, the *Chicago Tribune* said he had turned the Tennessee trip into "a rear platform speaking tour" as he spoke to nearly 50,000 people in one day. This was the man, according to the *Tribune*, who dominated the national Democratic Party for nearly three decades.[5] According to the *Washington Post*, "It would be difficult to find a man in American life today who had more staunch friends and more bitter enemies than William Jennings Bryan. . . . His voice has been heard, probably, by more people than any man on earth."[6]

Darrow's commitment to the working class came not from the campaign platform but via the bar. Before the Scopes trial, Darrow's reputation was built largely on his advocacy for labor and opposition to the death penalty. His defense of Eugene Debs and other American Railway Union officials in 1894 was his first big labor case. Darrow resigned from the Chicago & Northwestern Railway Company to defend the workers, who were arrested for defying an injunction that ordered strikers back to work. Though Debs and the others eventually were found guilty of defying the order, Darrow's reputation as a labor lawyer soared in the next decade. In 1902–1903, he represented striking miners before a presidentially appointed coal commission, winning an eight-hour workday, a 10 percent pay raise, and recognition of the United Mine Workers. The trial of William "Big Bill" Haywood, the secretary-treasurer of the Western Federation of Miners, was a national sensation. A bomb killed the Idaho governor at his Coeur d'Alene home after he requested federal troops to help put down a miners' strike. In July 1907, a jury found Haywood and one other union official not guilty, and was unable to reach a verdict on the third defendant.[7] In the defense of the

McNamara brothers in the *Los Angeles Times* bombing, Darrow was again defending labor.

His most famous case prior to the Scopes trial did not involve labor, but the death penalty. In the 1924 Loeb-Leopold trial, two well-to-do youths were charged with murder in the thrill-killing of a 14-year-old. Nathan Leopold Jr., 18, and Richard Loeb, 17, had confessed to kidnapping Robert "Bobby" Franks. The defendants' wealthy families hired Darrow, who withdrew not-guilty pleas in an attempt to save them from the death penalty. Darrow put his determinist philosophy to work in the trial. He rejected free will, good and evil, and any moral absolutes. Free will, he said, was a "barbarous belief" that spawned cruelty and wrongs. He said it was a fiction because people were not free agents in deciding their own courses in life.[8] Loeb and Leopold, Darrow argued, committed the crime because of forces beyond their own control. Specifically, he cited the inherent twists in their personalities and their environments, neither of which they could choose and both highly significant in determining the young men's actions. He apparently won over the court, which sentenced them to life in prison plus 99 years. Darrow's victory was more than saving them from the death penalty. It was a triumph for his unqualified opposition to capital punishment. In this case, Darrow showed that he meant not just to defend the downtrodden against injustice, but the privileged, too. His personal conviction stood true even in the face of such a heinous killing.

Radical Jeffersonian

Bryan, "The Great Commoner," offered himself as a man of the people, a true Jeffersonian. He believed deeply and sincerely in democracy, the people, and agrarian values. This was more than political expediency for Bryan. This was the essence of his world view, which also colored his perspectives on religion and science. He was part of an American-Protestant tradition that recognized the right of even the least educated to be heard, to rebel against not just theological hegemony, but intellectual hegemony as well.[9] Such a "declaration of rights" for the uneducated could move to full-blown anti-intellectualism and antielitism. Bryan's brand of antielitism and populist politics meant that if people opposed an idea, such as evolution, then they could veto it. At the least, they could cast it out of public institutions, such as schools. The reformer-politician Bryan used politics to improve social and economic conditions for people. The reformer-evangelist Bryan used religion to better the spiritual condition of individuals and the moral fabric of society. Religion and politics were, respectively, the manual and the lever for improving individuals and the social order.

Bryan's reputation as defender of agrarian America is well deserved and is classicly Jeffersonian—even to the extent of distrusting urban America and citing the small farmer as the exemplar of democratic values. Bryan repeatedly invoked the will of the people in his reform campaigns: farmers as the hope and foundation of democracy, and cities as a menace not only to liberty but to a moral social order. He expressed that genuine belief in the people in his 1894 loss in the U.S. Senate election. He wrote in the *Omaha World-Herald*, paraphrasing the *Book of Job*, "The people gave and the people have taken away, blessed be the name of the people."[10] In spite of being a habitual politician, Bryan probably was quite sincere when, in the wake of his glory days, he still extolled faith in common people and democracy. He stated in his *Memoirs* that one of his purposes in writing the book was "to show the goodness of the American people . . . the virtues that not only make popular government possible but insure its success."[11]

Scientific knowledge was not exempt from his radical majoritarianism. For Bryan, teaching evolution in public schools was more an issue of democracy than of science or religion. In an address prepared the month before the Scopes trial, Bryan identified four possible "sources of control" for school curricula. Elected legislatures, he said, would be the "natural sources" of control because "governments derive their just powers from the consent of the governed." He acknowledged local school boards but passed on those bodies because legislatures defined the locals' duties and elections. Scientists were not acceptable because they were a minority—no matter their expertise. He figured there to be about one scientist for every 10,000 people, based on his estimate drawn from membership in the "American Society for the Advancement of Science"[12]—"a pretty little oligarchy to put in control of the education of all the children." He rejected teachers as the curriculum authority because, he said, the public employs the teacher, who take direction from the employer. "[A] teacher receiving pay in dollars which is stamped 'In God We Trust,' should not be permitted to teach the children that there is no God. . . . That is the Tennessee case."[13]

His constituents were amenable to mixing Jeffersonian principles with Christian traditions as they lined up with him in opposition to challengers of traditions both secular and sacred.[14] In this way, Bryan and his followers could be simultaneously rebels and reactionaries, defying the establishment and defending tradition. He smoothly melded religion and politics with orations that almost always invoked Christian and biblical themes. Bryan described the 1896 presidential campaign as a fight for the nation's soul, as he assailed the Republican pharaohs and even claimed God's support on the silver standard. One historian disapprovingly gave Bryan credit for initiating a "new political culture full of programs and crusades." In his 1900 campaign, again against McKinley, Bryan repeatedly invoked both Jefferson and Jesus. While condemning American imperialism in the Philippines, for example, he likened the nation's actions to those of the British monarchy in 1776. He said it was not an issue of American power but the moral dictate of doing what was right. The "Cross of Gold" speech was a masterpiece of political-religious oratory, or perhaps religious-political oratory. He said, among many other things, "I come to speak to you in defense of a cause as holy as the cause of liberty—the cause of humanity." He properly is remembered for a phrase that like no other imbued a single political-economic issue with Christian righteousness: "you shall not crucify mankind upon a cross of gold."[15]

Public Intellectual

In his writing and lectures, Darrow shunned Bryanesque pandering to folk wisdom, and instead admonished the underclass, whom he often represented, to improve its lot through learning. English historian H.G. Wells called Darrow a "sentimental anarchist," deeming Darrow's idea of the common man "superstition": "He is for an imaginary 'little man'—against monopoly, against rule, against law, any law." Darrow's avid defense of individual rights fused with his personal philosophy and historic cases, particularly in defense of labor. In contrast to Bryan, Darrow believed Scripture was a byproduct of civilization, not its cornerstone. Darrow also was a contrarian by virtue of his socialist sympathies. The exaggerated inequities of Gilded Age capitalism sharpened his sense of injustice and his determination to fight those who exploited others. Socialism, he believed, was a logical system because it meant men themselves—versus some theologically evasive construct called "destiny"—were responsible for making the world a better place through mutual effort.[16] This complemented his hostility toward any religion that would invoke "god's will" or "destiny" rather than accept personal responsibility for one's own plight or for his fellow man.[17]

Darrow aspired briefly to be a professional public intellectual but found it was not profitable. Darrow's trial in 1912 for bribing a juror in the McNamaras' case injured his law career. Upon returning to Chicago, his greater ambition was to become a professional lecturer and writer. He still had a law firm, but it had no income, and so he traveled the public-intellectual route for a few years. He did not have the social impact that he envisioned, so in 1914 he went back to law, though he always remained in demand as a speaker.[18]

While Bryan mixed with the masses, Darrow preferred the company of the literati. Though an ardent defender of the masses, Darrow found them a stupid, prejudiced, and hateful rabble.[19] In Chicago, he resided among the intellectual elite. The poet Carl Sandburg said Darrow was "somewhat to Chicago what Diogenes was to Athens."[20] The "city of the big shoulders" that so charmed the poet was a good fit for the attorney. In contrast to his clientele, who often inhabited the city's grimier side, Darrow's dinner guests included editors and writers, scientists and artists, the likes of Sinclair Lewis, Sandburg, Theodore Dreiser, Charles and Mary Beard, and, of course, labor leaders.[21] Darrow's cultural life included the "Biology Club," a group of about forty men that was formed in 1915 to hear lectures on wide-ranging topics, including biology, geology, psychology, anthropology, and even biblical history. The meetings often were at Darrow's residence, only a few blocks from the University of Chicago, with UC faculty, businessmen, and professionals in the mix—"philosophers and near philosophers," in Darrow's words.[22]

The public intellectual is inevitably a minority position, which would not have been a problem for Darrow. His commitment to individual rights had broad appeal. His unequivocal opposition to capital punishment drew a mixed response. His agnosticism put him in a distinct minority. Darrow was a cultural rarity in his ability to condemn America's anti-intellectual impulses while wooing his audiences with folksy charm—a combination of talents almost never seen on the scientific side of the evolution debate since Darrow. He was eloquent, urbane, liberal, and edgy, a combination of virtues oft pursued and rarely attained. He could be a righteous scold who appealed to the highest learning in defense of the lowest man. The sharp edge of Darrow's public persona has been dulled by the legend given life in *Inherit the Wind*. Darrow's hard-line agnosticism was lost in the final scene of him, played by Spencer Tracy in the movie, thoughtfully balancing copies of *Origin* and the *Bible* in each hand as he exited the courtroom. And that after admonishing the Mencken character for deriding Bryan's beliefs as "fool religion": "You smart aleck! You have no more right to spit on his religion than you have a right to spit on my religion! Or lack of it!" Darrow would have agreed with the idea of freedom of expression, but it is highly improbable that he would have defended Bryan against Mencken in such a context.

The Evangelist and the Agnostic

Bryan was in the tradition of a nineteenth-century evangelical movement that, while not endorsing state-supported churches, advocated a church-state relationship. Like others in this tradition, Bryan saw the state as a tool for promoting moral values in society. A major figure in the movement, Lyman Beecher, found wide support for his condemnation of what he called "political atheism," which was shorthand for the idea that people, being inherently wicked, needed the guidance of a moral society.[23] Beecher believed society positively influenced individuals, restrained their darker impulses, and made them fit to live together. In the Preface to his *Memoirs*, Bryan summarized this marriage of the gospel and politics in a sentence concerning his accomplishments. He acknowledged a debt to those who helped him in the "spread of the Christian religion, the safeguarding of society, and the establishing of popular government."[24] It also helped explain why he seized upon social Darwinism as the great evil of the twentieth century. Social Darwinism would shackle moral constraints, he believed, and turn society into a contest of "survival of the fittest." Then, social order would collapse.

In the decade following World War I, Bryan combined populist appeal, antimodernism/anti-Darwinism, and press acumen. Here was the foundation of a political movement: constituents, cause, connections. Evolution was as much a target of convenience as of conviction. His moderation on this apparently immoderate issue was shown in the 1924 resolution he pushed the Florida legislature to approve. Cast to appeal to a wide swath of the public, it only condemned teaching evolution as a "fact," and, since it was not law, carried no penalty for teachers who did so. The resolution simplified the distinctions among atheism, agnosticism, and Darwinism by equating them and damning them: "[I]t is improper and subversive to the best interest of the people of this State for any professor, teacher or instructor in the public schools and colleges . . . to teach or permit to be taught atheism or agnosticism or to teach AS TRUE Darwinism or any other hypothesis that links man in blood relationship to any other form of life" [emphasis in original].[25] When newspapers took note of the resolution, Bryan took note of the attention and pressed his case for the resolution as a model for other states.

With his assumption of antievolutionism's national leadership in the early 1920s, Bryan's objection to evolution became even more politically pragmatic. Evolution provided a well-defined target—something immediate that engaged a large number of people. People could understand evolution's implications, and they could relate to it, especially if they had children in public schools.[26] In a 1924 Nashville speech, Bryan seemed to do nothing less than declare war on modernism when he stated that modernists were anti-Christian evolutionists.[27] He defined the relationship between Christianity and evolution starkly—"either-or," without middle ground. "It is time for the spiritual forces of the nation and the world to unite in opposing the teaching of evolution as fact," he stated. "[A]ll who give a spiritual interpretation to life are vitally interested in combating materialistic influences and in defending belief in God, the foundation of all religious faith." He declared the future of humanity at stake.[28]

Bryan's political star had faded somewhat after the 1908 presidential campaign against William Howard Taft. It was Bryan's third loss in a run for the office. He said he would not be trying again, but he did not abandon politics. He was shifting from Democratic Party leadership to being head of a religious movement. In either role, he was what is now called a media celebrity. As a Christian celebrity, he was not unique. It was a growing business at the time. One magazine-article writer called Southern revivalists "as thick and as thorough as crows in a Middle Tennessee cornfield."[29] However, Bryan was extraordinary in his combination of sense of audience and the press, ease with the public, and political skill. As he devoted more time to religious work, he took up the issue of evolution because he had heard numerous anecdotes of college students losing faith after accepting evolution, according to his widow, who assisted in writing his posthumously published *Memoirs*. He became convinced, she wrote, that "teaching of evolution as a fact instead of a theory caused the students to lose faith in the Bible."[30]

He became a polarizing figure, and "sides grew distinct," according to those *Memoirs*, whenever he took up a question.[31] It was not a new observation about Bryan. In 1923, a *Chicago Daily Tribune* article reported on Bryan's run to become moderator of the Presbyterian general assembly, meeting that May in Indianapolis. From within the assembly, one of the arguments against Bryan, according to the newspaper, was a "desire to avoid theological controversy, which it is alleged is certain if Mr. Bryan is elected." Bryan lost the race and then declined an appointment as vice moderator in order to devote more time to leading his antievolution campaign.[32]

By contrast, as early as the 1890s in public lectures, Darrow condemned Christianity as a "slave religion." He routinely put religion in diametric opposition to learning and education, and characterized the postwar antievolution campaign as bigotry and ignorance. In short, the Bible and evolution were irreconcilable.[33] For Darrow, the Scopes trial was another piece of evidence in his indictment of religion. Darrow believed the line between prophesy and madness was dangerously thin: "The number of people on the borderline of insanity in a big country is simply appalling. . . . Perhaps nothing so contributes to this as religious controversy."[34]

He scoffed that insanity and religious exaltation were different only in matters of degree: "If a man says he is living with the spirits today, he is insane. If he says Jacob did, he is all right. That is the only difference."[35]

Darrow believed one had to ignore learning in order to resort to God's will as an explanation for anything. Too often, religion was mere superstition in his opinion and any thinking person was an agnostic about something: "[O]therwise he must believe that he is possessed of all knowledge. And the proper place for such a person is in the madhouse or the home for the feeble-minded."[36] Those who championed the Bible as a guide to science roused special contempt from Darrow.

Darrow's insistence on an individual's right to challenge convention—whether in theology or social philosophy—muted his agnosticism among the general public. In this respect, he was able to take a risky, unpopular position, which was antireligion, and fit it to broader American cultural values of individualism and rebellion. Darrow's insistence that God was nothing more than a convenient fiction assured him minority status. And, thus, forever the rebel, forever the underdog.

Darwinism

Bryan was not antiscience, just anti-Darwin. Evolution was only one part of the modernist movement, but evolution was the part that most clearly collided with Christian tradition, at least as interpreted by some. Bryan, however, saw Darwinism as not just denial of biblical inerrancy, but of democracy itself, something that would render the idea of a common good outdated and irrelevant. He advocated science and technology, and to prove it even joined the American Association for the Advancement of Science in 1924. The timing suggests he may have been motivated to sign up for appearance's sake—covering himself politically rather than expressing genuine commitment. He believed scientific research could not ignore supernatural explanations because science then became materialistic, which for him inevitably meant immorality and rejection of Christian principles such as caring for the weak.[37] Bryan defined fundamentalism as literalism and antimodernism—Darwinism being exhibit number one among modernists' transgressions.

In a 1920 address in Washington, D.C., to the World Brotherhood Congress, Bryan said "Darwinian theory" was "the most paralyzing influence with which civilization has had to contend during the last century." He cited Frederick Nietzsche as the greatest Darwinian of all, and the one responsible for the philosophic foundation of German militarism. Nietzsche took "Darwinian theory to its logical conclusions, and died in an insane asylum." Darwinism, Bryan charged, turned democracy into a weakling's refuge. The theory "endeavored to substitute the worship of the superman for the worship of Jehovah" and inevitably resulted in war.[38]

An idea can't be brought to answer for its sins, but its advocates can. So Bryan went after teachers, especially those in eastern colleges and universities. Again, his political instincts were good because many of Bryan's constituents already resented university faculties, and most of his followers would not have known university faculty anyway.[39] In addition, his politics tended to be regional, stronger in the South, where colleges and universities were fewer. His base was weaker in the higher-education corridors of the Northeast. His support in newspapers reflected this growing regional identity of fundamentalism. Urban, East Coast newspapers in New York, Brooklyn, Philadelphia, and Boston were inclined to oppose him. Most Southern papers supported him, though it was sometimes more a habit of opposing Republicans than supporting Bryan.[40] His obituary in the *Boston Daily Globe* said he was surprised that "the Southern press had not taken cognizance of the criticisms by representatives of the Eastern press at the Scopes trial at Dayton."[41] He had grown to personify an antagonism between the rural South and urban Northeast, not just in politics but in religion, which eventually meant education and science. When Bryan took up the antievolution cause, four of the five states that either banned or condemned teaching evolution were Southern—Arkansas, Florida, Mississippi, and Tennessee. The remaining state, Oklahoma, had a large population of Southern immigrants. Though Bryan generally expressed his opposition in good, centrist, political fashion, his followers were less constrained.[42]

As for the campaign against teaching evolution, the journalist Walter Lippmann conceded that Bryan's logic was grounded in Jefferson. In the Virginia Statute for Religious Freedom, Jefferson wrote that it was wrong to compel people to support a religion or church or doctrine that they did not believe. Similarly, Lippmann wrote, Bryan had a valid point in arguing that people were being asked to support something they did not believe.[43] The concession, though, did not acknowledge any difference in the nature of truth—scientific truth via empirical research versus knowledge via faith. But compulsory support—via taxation—opened public schools as the battleground, making the issue relevant everywhere.

Bryan did not mean to exclude evolution from public schools so much as he aimed to put it in what he saw as the proper hierarchy of knowledge. The truth of science was part of the greater Truth of Scripture. Such thinking annoyed Darrow, who challenged Bryan in the pages of the *Chicago Daily Tribune* in July 1923 to answer a series of questions about the literal truth of the Bible. It was a preview to what would come a few years later in Tennessee when Darrow put Bryan on the stand to defend literalism. The *Daily Tribune* exchange also was a synopsis of opposing views of religion. Darrow accused Bryan of trying to "shut out the teaching of science from public schools." He said Bryan's answers to a few questions about the literal truth of the Bible "might serve the interests of reaching the truth." That idea—the nature of truth—was at the core of their irreconcilable worldviews of science and religion. For Darrow, religion was subject to empirical analysis and not exempt from the standards of evidence to which all other fields of inquiry were held. For Bryan, religion was the beginning of inquiry. In the *Daily Tribune*, Darrow asked, among other things:

> Do you believe in the literal interpretation of the whole Bible? . . .
> Was earth made in six literal days, measured by the revolution of the earth on its axis? . . .
> Was Eve literally made from the rib of Adam? . . .
> Did [Noah] build the ark and gather two pairs of all animals on the earth and the food and water necessary to preserve them? . . .
> Under the biblical chronology, was not the earth created less than 6,000 years ago? . . .
> Questions might be extended indefinitely, but a specific answer to these might make it clear what one must believe to be a "fundamentalist."[44]

In a response the next day, Bryan dismissed the questions as generated by a nonbeliever. The questions were irrelevant, according to Bryan, so he need not answer them. Instead, he assailed "theistic evolutionists":

> The man who denies the existence of God is not likely to have much influence, because evidences of a creator are so plain and innumerable that atheism when avowed is not nearly so dangerous as so-called theistic evolution.
> . . . It is an anaesthetic which deadens pain while religion is being removed.
> . . .
> I decline to turn aside to enter into controversy with those who reject the Bible as Mr. Darrow does.[45]

Even though Bryan deflected the issue from Darrow's questions about the validity of literalism, the response reflected the fundamentalist-modernist divide. In *The Story of My Life*, published several years later, Darrow illustrated this division in a quip he recalled making at the beginning of the Scopes trial. As the court was being called to order, Judge John Raulston "had a palmleaf fan in one hand, and the other the Bible and the statutes. As he laid these down on his desk, I wondered why he thought he would need the statutes."[46] In a way, Darrow was right because Raulston implicitly was weighing the truth of law against the Truth of Scripture. Darrow also understood that fundamentalists would not submit the Bible to critical scrutiny. "The fundamentalists denied that these Bible stories are legends, opinions, poems, myths and guesses, and pronounced them history. . . . The books of the Old and New Testaments were written ages before the world had any knowledge of our science."[47] At the Scopes trial, Darrow said, "Mr. Bryan was the logical man to prosecute the case. . . . [H]e did not represent a real case; he represented religion. . . . As to the science, his mind was an utter blank."[48]

When Bryan began his assault on higher education and the teaching of evolution in 1920, he returned in numerous speeches to the story line of an innocent young person corrupted in the university. Professors indoctrinated susceptible youth into materialist philosophy, as the students' faith faded. And the finale: "And this is done in schools and colleges where the Bible cannot be taught, but where infidelity, agnosticism, and atheism are taught in the name of science and philosophy."[49] His 1921 lecture tour, *The Menace of Darwinism*, was superficial and even incorrect in its conclusions about Darwin. But Bryan attracted huge crowds, and *Menace* became a defining document for antievolutionists. It first was delivered in the spring of 1921 and then reprinted as a pamphlet and newspaper column. Bryan eventually included it in his own volume on evolution, *In His Image* (1922), a collection that grew out of a lecture series at a Presbyterian seminary in Richmond, Virginia.

In His Image sold more than 100,000 copies, combining science, religion, and politics in a defense of the Genesis creation account and the Social Gospel. Bryan interpreted the latter as applied Christianity, placing "human rights before property rights" and putting Christian principles to work in creating a just and harmonious society. One of the lectures, "The Origin of Man," lambasted Darwinism for promoting immorality. Bryan said Christianity was neither unreasonable nor unscientific. He reiterated the charge that World War I showed the consequences of embracing Darwinism and abandoning Christ. Such a course also would violate the Social Gospel by culminating in a society that warred with itself by outlawing procreation among the less fit—the impoverished, handicapped, or mentally ill.[50] The accomplishment of the lecture series and book was not as a scientific critique, which they were not, but as mass appeal. The prose was clear and compelling, and the issue clearly defined. It was erroneous on multiple counts, such as failing to distinguish between social Darwinism and Darwinism, and not understanding the difference between the fact of evolution and the theory of natural selection. However, the volume provided a thin but accessible introduction to the subject and gave people a rationale for renouncing Darwinism.[51]

Bryan espoused faith as the foundation of education and knowledge. Darrow said he and the defense team entered the case "lest our public schools should be imperiled with a fanaticism founded on ignorance."[52]

The Scopes Trial

By 1925, Bryan's campaign was getting press attention, drawing crowds, and provoking legislative responses. He had excised from the issue anything resembling an intellectual debate and had confined the interests of religious America to a narrow set of objections that basically set evolution and religion in opposition. State legislatures would be the instigators, schools the battleground, courts the arbiters. The trial was a natural, if unintended, next step in Bryan's national campaign. To Bryan, Dayton, Tennessee, must have seemed an ideal setting.

Town leaders in Dayton thought the legal challenge to Tennessee's Butler Act would be a good chance to publicize their community and drum up some business. The law had passed overwhelmingly in both the state house and senate, to little public notice. Gov. Austin Peay signed it into law, thinking it never would be applied. The Butler Act made it illegal to teach humanity's descent from a "lower order of animals" or to deny the biblical account of creation. Peay was generally a progressive governor, having spent money on schools, highways, hospitals, and prisons. He told a state senator that the Butler Act was absurd and should not have been passed by the legislature, but he needed the support of rural legislators. He also noted, publicly, that the bill would not interfere with anything being taught in public schools, nor would it jeopardize any teachers. The Dayton boosters saw a newspaper item about the American Civil Liberties Union offering to defend a test case against the law and thought it was an opportunity, something one of them said "would put Dayton on the map." The men who were planning the event figured Scopes, a science teacher at the local high school, to be a good choice to put on trial. He did not have a family, so others would not be injured by any unforeseen consequences. Scopes opposed the law but was not a vocal opponent. His opposition was a matter of his valuing education and freedom of inquiry. As a rather cautious agnostic, he was friendly to religion but never committed to it. The issue caught fire in Robinson's Drug Store during a series of discussions about evolution, the Butler Act, and promoting Dayton. Finally, those who initiated the idea of a show trial asked Scopes to join them. When he showed up at the drug store, Scopes said he did not know how it was possible to teach biology without teaching evolution. He reluctantly agreed to be arrested for the sake of civic boosterism.

The trial began July 12, lasted for twelve days, and ended with Scopes's conviction and a $100 fine. It was, as expected, a media frenzy with newspapers and wire services from across the nation showing up for the trial, which starred Darrow and Bryan. It even attracted radio coverage, WGN of Chicago. Only five days after the end of the trial, Bryan died in Dayton. He had been working on polishing an antievolution speech that would have been his closing argument had Darrow not waived closing statements. He had delivered parts of the speech to enthusiastic crowds in nearby Winchester and Jasper the day before his death. On Sunday, July 26, after leading a prayer at the Methodist Church, South, he returned to the private residence where he had been staying, ate lunch, took a nap, and died in his sleep. Shortly thereafter, efforts began to build a fundamentalist university as a memorial to Bryan. William Jennings Bryan University, later shortened to Bryan College, opened in 1930 in Dayton.[53]

The state supreme court overturned Scopes's conviction because the judge, rather than the jury, as required by law, had set the penalty. This robbed Darrow and the defense of the opportunity to take the case to the U.S. Supreme Court, which had been their goal. The Tennessee legislature repealed the Butler Act in 1967, but that did not end the fight over teaching evolution, because legislation has been introduced periodically to put Genesis back in the classroom or to diminish the veracity of evolution as science. In 1974, Tennessee legislators approved a bill requiring equal time for various theories of origins, which meant including Genesis. In 1996, the General Assembly attracted international attention with a bill that would prohibit teaching evolution as a "fact." The senate eventually defeated the bill. In 2012, the state legislature passed a bill that protected teachers who allowed students to challenge "controversial" scientific ideas such as evolution. It became law without the governor's signature.

Though it is easy to caricature Bryan as the blowhard simpleton of *Inherit the Wind*, that was not the case. He was a logical participant in the trial inasmuch as he was the leader of the national antievolution movement. He represented many Americans' views of the state and local prerogatives concerning education, and their concerns about the rising indifference to sacred traditions. Viewed as political opportunism, rather than a legal showdown against the era's greatest trial lawyer, Bryan's role makes much more sense. In this respect, he actually had an advantage over Darrow. Bryan was more politically accomplished. His impact on the Democratic Party had been profound. Bryan's argument about the immediate issue—teaching evolution in public schools—revolved around the rights of voters and their representatives to determine the education of their children in publicly funded institutions.

The addition of Darrow amplified the trial's celebrity factor, which is usually a good thing for politicians. Bryan's approach was to win a verdict much as one would win an election. This, as much as the issue itself, may have been his undoing in the trial. In what one historian called Bryan's "quest for dramatic settlement," the Great Commoner sought a showdown, which he compared to the one between Elijah and the prophets of Baal. Bryan challenged modernists—atheists and materialists—to write "a better Bible than ours. . . . use to the full every instrumentality that is employed in modern civilization, let them embody the results of their best intelligence in a book and offer it to the world." It was mere speechmaking, of course, because no one in the opposition was going to take him up. But it was good politics, aimed not at his nemeses, but at his constituents, who would hear the rhetorical gauntlet thrown down.[54]

Darrow and his friend H. L. Mencken shared, to varying degrees, contempt for Bryan, whom Darrow deemed the "idol of all of Morondom."[55] Darrow recognized and held in disdain Bryan's exalted status, in Mencken's words, as a "Fundamentalist Pope."[56] The ridicule of Bryan, even in death, served the immediate purposes of proevolutionists. It was easier to assail an individual than a system of beliefs that was inconsistent, ill defined, and scientifically illogical. Bryan, its flag-bearer, was a distinctive and public target. But the verbal barrage was inconsequential in the long-term battle, in which Bryan's defenders glorified his self-sacrifice and righteousness in defense of a holy cause. Both Darrow and Mencken believed Bryan a fool, but not a harmless one. Again, as usual, Mencken said it best: "He seemed preposterous, and hence harmless. But all the while he was busy among his old lieges, preparing for a jacquerie that should floor all his enemies at one blow. He did the job competently. Heave an egg out of a Pullman window, and you will hit a Fundamentalist almost anywhere in the United States today. . . . They are everywhere that learning is too heavy a burden for mortal minds."[57]

Years earlier, Darrow raised doubts about Noah, the flood, the ark, about Joshua making the sun stand still, and other Scriptural stories. He detailed his skepticism in his 1910 volume, *Clarence Darrow on Religion: Facing Life Fearlessly, Absurdities of the Bible, Why I Am an Agnostic*.[58] As noted earlier, the courtroom exchange also had been rehearsed in the *Chicago Daily Tribune* two years earlier. The trial was not a new drama, just a bigger theater.

The trial appealed to Darrow for several reasons. First, it was a chance to go after Bryan to even a political score. Darrow blamed him for the misfortunes of the Democratic Party as a result of Bryan's three defeats in presidential elections. Second, the trial was a chance to put fundamentalist Christianity on trial. As the evangelical politician of his day, Bryan was an irresistible target for a man who had spent his career criticizing religion. Third, Darrow believed in freedom of thought and expression and abhorred attempts to constrain individual liberty. Their fight became a contest between a militant agnostic and the spokesman for fundamentalist Christianity.

When Darrow put Bryan on the witness stand as a religion expert, it was the kind of face-off that Darrow wanted. He would challenge religion in general, biblical literalism in particular. The exchange was the hard-line rationalist against the unyielding man of faith. They were good figureheads for their respective sides, good performers, and good theater for the press, which was largely absent from the trial by the time of the now-mythic showdown moment. Bryan accused Darrow of trying to "slur . . . the Bible," to which Darrow responded that he was simply questioning Bryan about his "fool ideas."[59] That great stage for Bryan and Darrow was a spectacle—replete with "simian sodas" for refreshment, "Monkeyville police" to keep order, and street preachers exhorting, entertaining, and damning. Historians and filmmakers have judged Darrow the winner, even though he lost the case. The contrived event grew into a forum on science and religion, modernism and fundamentalism—charismatically presented by Darrow and Bryan.

Fundamentalists cast the debates in mythic terms that would be comprehensible to anyone, whether or not one agreed with the contentions. At issue for Bryan: equality and antielitism, individual choice, Davids and Goliaths, embarking on new frontiers, the Jeffersonian agrarian tradition against the menace of urbanism and—by implication—the foreigners who lived in the cities. It was natural for Bryan, the populist, to tap into antiurban, antielitist sentiments, which he had done in his presidential bids. Such egalitarianism fit neatly with democratic values, and so put science to the vote. Based on this perspective, a scientific theory could be validated by public approval, like a candidate for office. As for the assault on rural life, that was the defense team: Darrow, of Chicago; John R. Neal, of Knoxville, and a former law professor at the University of Tennessee; Dudley Field Malone, of New York; and Arthur Garfield Hays, of New York, and the sole ACLU representative on the team. The prosecution saw them as invaders, proof of urban aggression upon rural America and disdain for its traditions. Bryan's venomous antagonist, Mencken of the *Baltimore Sun*, confirmed it.[60]

The Scopes trial was important for Darrow as a public intellectual. Though he was the nation's village atheist throughout his career, he often played on a stage where his faith was no issue. Darrow admitted getting into the case in order to put a spotlight on Bryan's antievolution, antieducation campaign.

Only after learning Bryan was part of the prosecution did Darrow agree to join the defense.[61] In Dayton, Darrow basked in his contrarianism and agnosticism, in severe contrast to the rural community's norms of faith. Dayton amplified both Darrow's agnosticism and his rebellion. It was an ideal platform, simultaneously the picture of small-town America and a parody of its culture, like a toothily grinning "American Gothic."[62] In personal style, Darrow was the antithesis to Dayton. His law office in Chicago was "a Mecca for the lonely and the famous—the worker seeking legal advice, the intelligentsia looking for intellectual stimulation, the literati passing through Chicago, the prisoner just released from jail, the indicted facing trial, the lone black seeking help, the representative of a black organization asking the attorney to speak at a fund raising."[63] He was the defender of everyman and the hanging judge of common ignorance.

Bryan's *Last Message* and Darrow's Opening Argument

What would have been Bryan's closing argument to the jury, his posthumously published *Last Message*, pulled together many of his core ideas—the values to which he appealed, the style of argument, and the approach to science itself. It was never delivered to the jury because Darrow and the defense declined to present a closing argument, thereby robbing the prosecution and Bryan of the opportunity to do so. In his *Last Message*, Bryan began by praising the common man, then castigated minorities or "cults" whose beliefs about science were at odds with the Great Commoner's and his followers. At the outset, he praised the "sturdy honesty and independence of those who come into daily contact with the earth. . . . I admire the stern virtues, the vigilance and the patriotism of the class from which the jury is drawn." He argued not about science or religion, but majority opinion: "The majority is not trying to establish a religion or to teach it—it is trying to protect itself from the effort of an insolent minority to force irreligion upon the children under the guise of teaching science. What right has a little irresponsible oligarchy of self-styled 'intellectuals' to demand control of the schools of the United States, in which 25,000,000 of children are being educated at an annual expense of nearly $2,000,000,000?"[64]

In those few sentences, Bryan artfully defined the issue as one of democratic rights in the face of an elite "oligarchy" poised to seize control of schools. It was more than just the majority he was defending, he said later in the address. He was fighting a defender of the rich and the immoral: "Mr. Darrow . . . was engaged about a year ago in defending two rich men's sons who were on trial for as dastardly a murder as was ever committed. The older one, 'Babe Leopold,' was a brilliant student, 19 years old. He was an evolutionist and an atheist."[65] Via the false syllogism, Bryan had folded intellectualism neatly into evolution, atheism, and murder. Bryan implicitly contrasted his "yeomen" with those wealthy, educated, urban youth and their deeds.

The will of the majority was one of the twin spires of his belief system. The other was the authority of the Bible. Bryan understood Americans' adoration of science and technology as measures of progress. So he did not slight them. Instead, he praised science as being "of incalculable service to man," and cited electricity, steam power, telephones, radios, sewing machines, harvesters, automobiles, and even "artificial ice." The last item may have been close to the hearts of all in that hot July day in Tennessee. Medicine was of "invaluable service." But all of these things remained beholden to the highest truth: "Christianity welcomes truth from whatever source it comes, and is not afraid that any real truth from any source can interfere with the divine truth that comes by inspiration from God Himself. . . . Evolution is not truth."[66]

And so the case was closed, as far as Bryan was concerned. Because evolutionists failed to defer to Scriptural truth, their science failed. This made literalism, then and now, irreconcilable with "materialist" science. For Bryan and his successors, science was fine, as long as it recognized biblical authority as the ultimate test of truth. Those who did not understand this, according to Bryan, were a minority, a few cultists. At the close of the argument, Bryan declared the necessary role for the "Christian state": "Again, force and love meet face to face, and the question, 'What shall I do with Jesus?' must be answered. A bloody, brutal doctrine—Evolution—demands, as the rabble did nineteen hundred years ago, that He be crucified. That cannot be the answer of this jury representing a Christian state."[67]

Majority will and biblical authority were his bookends of truth. In all, *Last Message* was a masterful document—not of science, or of theology, or of law. It was a campaign speech, with the immediate purpose of winning the jury votes. But its larger target was the national audience. It was a tactical roadmap of ideas and values to enlist against evolution, in the same spirit that he had used a few decades earlier against trusts, railroads, the gold standard, and Republicans.

Darrow's opening argument was the trial's intellectual high point. The questions to Bryan were headline-grabbing ridicule, personal attacks on Bryan and the ideas he represented. The opening statement was the freethinker's exegesis on religion, particularly state-supported religion, cast before an unsympathetic audience, both locally and nationally. Darrow did not invent the ideas, but his celebrity, eloquence and intellect gave them force. In particular, he concentrated on two things. The first was the enduring war between science and religion, which Darrow made surrogates for enlightenment and ignorance, tolerance and intolerance. The second was individual rights, including the right to dissent, even to disdain the sacred. Darrow told the court—and America—that religion was divisive and cruel at worst, simply antieducation at best. It was, in his words, medievalism, "the setting of man against man and creed against creed until with flying banners and the beating of drums we are marching backward to the glorious ages of the sixteenth century when bigots lighted fagots to burn the men who dared to bring any intelligence and enlightenment and culture to the human mind."

While he seared religion for being the foremost cause of "bitterness, of hatred, of war, of cruelty," he asserted it was something "that should be left solely between the individual and his Maker, or his God, or whatever takes expression with him, and it is no one else's concern."[68] In his plea for tolerance, Darrow challenged the right of the state to legislate faith and defended the right of the individual to believe what he or she may: "I know there are millions of people in the world who look on it [the Bible] as being a divine book, and I have not the slightest objection to it. I know there are millions of people in the world who derive consolation in their times of trouble and solace in times of distress from the Bible. I would be pretty near the last one in the world to do anything or take any action to take it away."[69] However, he denounced using the Bible as a "yard stick to measure every man's intelligence and to measure every man's learning":

> Are your mathematics good? Turn to I Elijah ii, is your philosophy good? See Samuel iii, is your astronomy good? See Genesis, Chapter 2, Verse 7, is your chemistry good? See—chemistry, see Deuteronomy iii–6, or anything that tells about brimstone. Every bit of knowledge that the mind has, must be submitted to a religious test. . . . If men are not tolerant, if men cannot respect each other's opinions, if men cannot live and let live, then no man's life is safe, no man's life is safe.[70]

While he spoke, the *New York Times* reported, the courtroom was quiet. The *Chicago Tribune* wrote that it was one of the greatest speeches of Darrow's career. The news inferno erupted, with more than 200,000 words surging over telegraph lines that day, eclipsing any other national event for coverage.[71] Newspapers quoted him at length, often admiringly, but probably to little effect in changing minds about science and religion, and only reinforcing existing inclinations. In the course of defending Scopes—or, perhaps more accurately, prosecuting religious tyranny—Darrow polished themes that have been resurrected with each new episode of the debate, especially individual rights, condemnation of intolerance, and the irreconcilability of science and religion.

The Two Truths

Arthur Pierson, who was prominent in the American evangelical movement of the late nineteenth century, wrote in *Many Infallible Proofs* (1895): "I like Biblical theology that . . . does not begin with an hypothesis and then wraps the facts and the philosophy to fit the crook of our dogma, but a Baconian system, which first gathers the teaching of the word of God, and then seeks to deduce some general law upon which the facts can be arranged." The appeal to Francis Bacon, the seventeenth century advocate of objective, empirical inquiry, would have resonated with many Americans in the nineteenth century. Evangelistic Baconianism meant that one already had at hand the hard "facts" in the Scripture. The challenge was to organize them in a way so as to reveal the greater pattern, or Truth.

The "inerrancy" of fundamentalist thought grew out of the idea that facts— i.e., the Scripture—could not be wrong. The facts were inerrant because they were dictated directly by God rather than interpreted by human agents with the filter of language, not to mention translation into other languages. The facts revealed in the Bible were, for fundamentalists, just as factual as the physical laws described by Newton. It was up to people to discover the meanings, inherently true and factual, in the Bible. This left the interpreter to seek out clues and arrange them in order to come to a greater understanding of Truth. Thus, the researcher did not come to the facts with a hypothesis to be tested, but instead to sort, classify, and reach general conclusions. In this tradition, fundamentalists adopted a single model of interpretation. Since Newton and Bacon, new scientific discoveries had been incorporated into old beliefs as new scientific knowledge was interpreted to be more evidence for design in nature. For Bryan, this meant he could be both literalist and scientific. He could claim with sincerity that true scientists, with this Baconian approach, did not impose an order—an hypothesis—on the facts, which were inviolable by virtue of their supernatural inspiration. Instead, to Bryan, true scientists sought order and understanding from the facts themselves, which were Scriptural as well as of the natural world.[72]

Darwin, though, in a philosophical tradition developed by Immanuel Kant and others, redefined the relationship of science and religion. Darrow and modernists cast off the Baconian approach, the single-model view of understanding facts. Instead, they came to the facts in a Kantian fashion, which allowed multiple perspectives and speculation. The perception of fact became part of a process. Science was autonomous and separate because it was in the realm of things known by the senses—things experienced directly. Religion was the realm of spirituality and morality and not amenable to empirical inquiry. Science was no longer subject to religion.[73] Darrow and the modernists' rejection of fundamentalism, then, would not be so much a matter of a priori denial as it would be a matter of religion being the least empirically valid approach to the facts.

Darrow and Bryan were among the most publicly visible representatives of these opposing philosophies in the early twentieth century. In the same way that Kantian and Baconian traditions were incompatible, modernism and fundamentalism were incompatible, Bryan and Darrow were irreconcilable. For Bryan, the Darwinian idea that random chance was at the center of an obviously ordered nature was simply a self-evident contradiction. For Darrow, it was equally self-evident that Scriptural validity should be subjected to the same critical scrutiny as any other text or set of facts. That wall of incompatibility still stands.

3

From the Scopes Trial to *Darwin on Trial*

If the enemy sends its Goliath into battle,
it magnifies our cause. . . . If St. George had slain
a dragonfly, who would remember him.
—Brady (Bryan character), from *Inherit the Wind*, Act 1

Bryan's death threw fundamentalism into some disarray.[1] The movement lost focus, but not energy, as creationists reorganized, established publications, and took advantage of new media platforms—radio in the 1930s and television in the 1950s. The loss of Bryan's leadership meant reorganizing and rethinking his progressive politics and liberal interpretation, by literalist standards, of scripture in which "days" could mean "ages." During the next few years, several individuals campaigned to become Bryan's successor. His friend George F. Washburn founded the Bible Crusaders of America to lead people "Back to Christ, the Bible, and the Constitution." Another evangelist, Paul Rood, in California, helped found the Bryan Bible League and claimed nothing less than a vision from God calling on him to succeed Bryan. The Defenders of Christian Faith, headed by Gerald Winrod of Kansas, sent squadrons of "flying fundamentalists" around the Midwest promoting antievolutionism. The Supreme Kingdom, founded by a former national organizer for the Ku Klux Klan, adopted Bryan's cause against evolution and modernism for its agenda.[2]

The groups did not last long, but their existence indicated a subtle shift in fundamentalism, which remained antievolution but began to look more like a political party than an evangelical tent revival. Advocates attempted to get the messages and tactics in concert across the country, organizing around central issues, reminding parishioners of the larger problems beyond the church doors. In 1929, Walter Lippmann noted fundamentalism's change to a political movement, as well as its oddities and irrationalities: "In actual practice, this movement has become entangled with all sorts of bizarre and barbarous agitations with the Ku Klux Klan, with fanatical prohibition, with the 'anti-evolution' laws and with much persecution and intolerance." Lippmann, one of the great journalists of the era, edited the *New York World*, a prominent and influential liberal voice. He wrote that fundamentalist fear of evolution was appropriate because if the gospel was only symbolic, not a historical record, then the foundation of religion was under threat. "The angry absurdities which the fundamentalists propound against 'evolution' are not often due to their confidence in the inspiration of the Bible. They are due to lack of confidence. . . . But because their whole field of consciousness is trembling with uncertainties they are in a state of fret and fuss." Lippmann said Bryan was logically consistent in arguing for majority veto of teaching evolution, but in the process Bryan reduced the idea of majority rule to an absurdity, a device for preventing education. The ultimate inanity of the idea lay in the presumption that the doctrine of spiritual equality meant all people were equally good biologists.[3] The alliances with political extremes and the embrace of logical absurdities preceded a decline in fundamentalists' numbers as more moderate, even conservative, Protestants simply dropped their support in order not to be associated with fundamentalism's more outlandish factions.[4]

Many educated people ridiculed fundamentalists, having adopted Mencken's farcical take on the Scopes spectacle. His caricature became the more urbane, elite shorthand for denigrating rural and small-town America. His contempt was nothing new—just more scathingly eloquent. Such disdain had been running through the period's literature for several years, in such works as Edgar Lee Masters' *Spoon River Anthology* (1915), Sherwood Anderson's *Winesburg, Ohio* (1919), and Sinclair Lewis's *Main Street* (1919). Mencken's brutal trial dispatches about fundamentalists amplified the cynicism, went to a wider audience, and added laugh lines. Lewis used religion in his 1927 novel *Elmer Gantry*, whose shyster protagonist invoked fundamentalist ideas in a campaign of moral reform. In 1929, Thomas Wolfe added viciousness to the list of small-town abominations in his novel *Look Homeward, Angel*. All those outside metro-America seemed to get jammed into Mencken's absurdist mural of life in the hinterlands among prayer-afflicted yokels. Fundamentalism had become popular parlance for much more than a Protestant faction. It meant, among other things, backward anti-intellectualism, a charge that remains into the twenty-first century. Another consequence of the caricature, especially Mencken's, was to stain fundamentalist arguments with obscurantism. Their views would not be taken seriously in the mainstream.[5]

Post-Scopes Fundamentalism

The Scopes trial may have seemed a triumph for modernists and science in the 1920s. However, the trial was anything but the defeat of antievolutionism. Many accepted the so-called "Scopes legend"—that antievolution was beaten with Darrow's relentless witness-stand grilling of Bryan about his "fool ideas." In fact, the cause had only retreated from the headlines. It was quite alive and well. About two dozen antievolution bills were debated in state legislatures from 1925 to 1929. In 1927, eighteen such statutes were introduced in fourteen states. Most failed, but the movement's proponents were undeterred. After an antievolution bill failed to move out of the Texas House of Representatives in 1929, it took thirty years for another antievolution bill to be introduced into a state legislature. Antievolution laws were still on the books in three states, including Tennessee, but prosecutors did not enforce them. Fundamentalists remained committed as followers began a decades-long campaign against evolution in public-school texts and curricula.[6]

The fundamentalist reorganization meant a shift away from single-person charismatic leadership to groups organized around their convictions. Creationist societies organized in the mid-1930s, such as the Religion and Science Association (RSA), were an attempt to establish a united front against evolution. Though it did not last long, the RSA demonstrated a new approach that meant organizing institutions with evangelical intent and tactics—spreading the word via higher education and radio while lobbying state legislatures and local school boards. Fundamentalist efforts went from state level to local level, and this approach worked rather well. Smaller communities were easier to pressure, and it was easier to identify those more amenable to the message. The press coverage declined, so fundamentalists began their own publishing operations, showing the lessons learned about the power of the press and public opinion.[7] In some cases, the publications had impressive circulations, such as the *Defender Magazine*, to which 600,000 people subscribed, and the *Christian Beacon*, with 120,000 subscribers. The *Defender Magazine*'s publisher was far-right conservative Gerald Winrod. In the 1920s he organized the Defenders of the Christian Faith, whose purpose was to fight the teaching of evolution. In the 1930s, he condemned Franklin Roosevelt's "Jewish New Deal" as satanic, this after traveling to Nazi Germany to study the "Jewish menace."[8] The religious message remained at the center of fundamentalism, but the message and messengers had taken on a decidedly political aura in terms of activism, organization, and media.

Antievolutionism was only one aspect of fundamentalism, but it was an important one—familiar to all, understood by few, and a topic that could draw a crowd. The shift in tactics meant not only becoming more political but co-opting the vocabulary of science. Earlier, Bryan had accused evolutionists of being unscientific, of offering only an unsupported "hypothesis" lacking in factual support. Publishing their own journals, establishing their own institutions of higher learning, learning the language of science, fundamentalists were tacitly acknowledging their vulnerabilities in their fight with evolution. As had been the case with the religious message, their science message was aimed at the nonexpert—those who knew *of* evolution, but little *about* it. The approach was nothing new as a political campaign: exploit concerns and fears; condemn the opposition for leading one to destruction; offer an alternative that assuages friends, damns foes, and does not require a lot of thinking.

Antievolutionists may have been forgotten on the front page, but they were at work creating a new institutional base of conferences, institutes, and journals. As the antievolution campaign began to focus on school boards and local government, the primary targets became schools, teachers, and textbook publishers. Though not as visible as earlier campaigns, it was effective, and evolution began to disappear from high-school texts. As many as one-third of teachers feared being labeled an evolutionist. According to a 1942 Carnegie Foundation survey, less than half of all high-school biology teachers even addressed the subject. The campaign still provoked contempt in many corners. Disdain in urban, East Coast America was illustrated in Delaware, where in 1927 a proposed ban on teaching human evolution in public schools was referred to the Committee on Fish, Game and Oysters. [9] But antievolutionism's power was not in urban America, and the movement's leadership suffered no illusions about where its support resided. As a Southern strategy emerged in the late 1920s, those individuals whom Mencken had derided as rubes became the core of a consciously populist campaign set against urbanism, elitism, and the erosion of traditional values.

At about this time, the term that came to define the cause—*creationism*—appears to have arisen. In 1929, Harold Clark published *Back to Creationism*, which advocated using the label "science of creationism." His ideas were unoriginal, a repackaging of concepts from his mentor, George McCready Price and his "flood geology," which meant a single act of creation, in a twenty-four-hour day, and denied other fundamentalist interpretations that might accommodate multiple creations and millions of years. Price was a Seventh-Day Adventist, largely self-taught and rather eccentric. His defense of the six-day creation, in *The New Geology* (1923), declared that scientists were simply wrong about the amount of time it took for the geologic strata to form. Rapid sedimentation and the fossil layers took not billions of years, but only a few years during a great catastrophe—Noah's flood. *New Geology* was the foundation for later creation-science. He wrote that no proof existed for an old Earth; that the so-called historical order of fossils was simply a mistake; no proof existed that dinosaurs and later mammals of the Tertiary era did not coexist; no one could show that the ancient, fossilized marine forms did not coexist with all other life forms.

Though Bryan accepted the day-age theory of creation, he liked Price's argument, as did many other scientifically uninformed people. Scientists largely ignored Price. In 1937, an ally of Price's began circulating *The Creationist*, a mimeographed sheet that endorsed broader readings of Genesis and included day-age interpretations. *Creationism* was—and is—a good term for several reasons. First, it was distinctive for this singular cause, nontechnical and easily comprehensible for an audience and for media consumption. Second, as with other political terms, such as *liberal* or *conservative*, the word *creationism* was value laden. It connoted piety and devotion to Christianity without degenerating into theological nuances and complexities. Third, it no longer was "anti" but was positive, the same sort of rhetorical capital gained when a group is not, for example, antiabortion but prolife.[10]

The antievolution movement found natural alliances in politics. Prohibition was one of them. The 1928 presidential election revealed the growing political acumen and power of fundamentalists. Democratic candidate Al Smith was an obvious target for fundamentalists because of his proevolution views. He also was a New Yorker who opposed prohibition, and he was Catholic. With the specter of "rum and Romanism" running America, fundamentalists joined prohibitionists against Smith. Some fundamentalists, however, saw such partisan activities as degrading church unity and integrity. Many more came to see the political victory as a hollow one, because dissention among the ranks sharpened, attendance at debates declined, and public policy remained essentially intact.[11] Fundamentalists were no longer grabbing headlines or ramming bills through legislatures. But antievolutionism remained central to fundamentalists' political activism and theological identity. More liberal Protestant denominations prevailed in the 1920s and 1930s over the fundamentalist movement, which splintered into small churches. Some became vituperative and judgmental. They still saw themselves as the true adherents to and guardians of American Christian culture. By 1940, they began to reemerge in public life, and this was most immediately evident in the creation of the American Council of Christian Churches. It was established as a counter to the Federal Council of Churches, which liberal denominations used to influence law and policy.[12]

Though still fighting modernism, fundamentalists learned to love some of its manifestations. One was radio, which in the 1920s had brought new ideas, new music, new mores from the city to small-town and rural America. What had been local or regional became national. As radio grew in the 1920s and 1930s, so did fundamentalist radio ministries. When fundamentalists took advantage of growing commercial radio in the 1930s, there was no sudden stream of antievolution broadcasting, but the trend signaled a growing audience for conservative Christianity. By the mid 1940s, the *Old Fashioned Revival Hour* was the most popular program on radio, carried by more than 450 stations nationwide. The rapid growth of religious radio in the 1920s helped launch fundamentalist and antievolutionist programming, which eventually reached audiences in the tens of millions.[13] Before the end of the 1940s, more than 1,600 fundamentalist programs were being aired weekly. Fundamentalists' use of the era's "new media" revealed some ambivalence on antievolutionists' part about the modern world. They rejected "modernism" per se, but understood the value of science and technology. This ambivalence was on national display when Bryan declared evolution unscientific because it was a mere "theory," meaning he accepted facts but rejected the empirical foundation of evolution. Similarly, fundamentalists rejected the implications of science for literalism, but rushed to the technologic fruits of science to spread their messages.

Fundamentalist campaigns moved from converting the mainstream to reinforcing the foundation. Their publishing enterprises expanded, along with Bible camps, foundations, schools, and colleges.[14] As fundamentalists retreated from the mainstream after the Scopes trial, they began to create their own culture within a culture. Bryan had blamed academe for a national decline in faith. He recounted in numerous articles and addresses the tales of young people returning home spiritually adrift, having lost faith as a result of the secularist education that put facts before faith. It was a natural response among fundamentalists to fix the problem by creating their own educational system. They were quite successful in this venture. By 1930, more than fifty fundamentalist colleges had opened, and another twenty-six were established during the depression years. During the 1930s, Wheaton College, in Illinois, was for several years the fastest growing liberal arts college in the nation.[15]

The evolution controversy cooled over the next few decades for the simple reason that more immediate problems appeared, primarily the Great Depression and World War II. This did not mean the disappearance of antievolution as much as a shuffling of priorities. The postwar world was scary—nuclear power, the Cold War, the Soviet Union touting its technologic prowess.

America had saved the world, but now was being challenged by communism, which denied core values in the American myth, especially exceptionalism, individualism, and America as the new Eden and new frontier. World War II was a sort of secular Armageddon, so many people believed it was time for the real one. The 1940s had shown the incredible, apparently unlimited, capacity for humanity's inhumanity. World War II was an opening for antiscience, in much the same way World War I had been a catalyst for antievolutionism. Some saw the atomic explosion as the precursor of the biblical apocalypse. *The Atomic Age and the Word of God*, published in 1948, was a best seller. The author, Wilbur Smith, argued that the atom bomb was proof that the literalists were correct, that the word of the Bible was inerrant. *Eternity* magazine declared in 1945 that the bomb showed the inevitable fulfillment of the divine plan, and there was no hope for avoiding a final, fiery judgment.[16] Antievolutionists persisted with publications and legal challenges to teaching evolution in schools. They continued to lose in court, a pattern that has held into the twenty-first century. But they won national attention in a series of notable cases, and they assumed the role of the aggrieved underdog, set upon by the establishment, rebels condemned in their quest for justice. The quiet triumph of antievolution was the absence of evolution in public-school biology classes. The subject had declined to the point of near disappearance before beginning to reappear in the late 1950s and early 1960s.[17]

1950s and 1960s: Modernism Gone Amok?

Religion grew in the 1950s, not just the conservative fundamentalism but religiosity in general. A 1957 survey by the U.S. Census Bureau showed an astounding 96 percent of respondents reporting an affiliation when asked, "What is your religion?" Even allowing for the inflation prompted by the wording of the question, it was still a big number. Pulpits increased, too, as church construction became the fourth largest private-building category. The religion boom was more than dollars and numbers. It was growing in prestige.[18] The historian Martin Marty wrote that "the religion of political fanaticism reached its peak with McCarthyism" in the first half of the decade. It was "theistic democracy" at war with atheistic communism, a conflict brought to the nation's living rooms by Sen. Joseph McCarthy, who owed much of his notoriety to television. This potential of television was not lost on others, including religious conservatives.[19] Again, many Americans found religion and politics perfectly congruent. Just as fundamentalists and creationists found that Darwin challenged the Old Testament, many people believed communism threatened Christian civilization.

Televangelist ministries, which grew into empires, boomed in the 1960s and 1970s, often as complements to megachurches. Jerry Falwell's *Old Time Gospel Hour* was estimated to have reached 40 percent of U.S. households. By the late 1980s, his services were broadcast over 392 television channels and 600 radio stations. The programs were moderate in tone, compared to the extravagant showmanship and outlandish admonitions of other fundamentalist ministries. The televangelists had a ready constituency in the South, where in the wake of the 1960s turbulence people again were feeling alienated and apart. Many turned to fundamentalist churches, including televangelists. The earliest audience and potential converts for the broadcast churches lived along the edges of the South, including Virginia Beach, Virginia, where Pat Robertson began his Christian Broadcasting Network and *700 Club*; Lynchburg, Virginia, where Falwell started his televangelist ministry; and Charlotte, North Carolina, the home ministry of the later scandal-ridden ministry of Jim and Tammy Faye Bakker. By 1979, according to a Gallup Poll, about 1,300 Christian radio and television stations had an audience of 130 million. Profits were estimated at between $500 million and billions.[20]

In the midst of the 1950s doctrinal turmoil among Protestants—liberals, conservatives, and moderate evangelicals—a charismatic leader with media savvy and appeal emerged. Billy Graham was well connected with conservative fundamentalists in the South, but was himself more moderate. As part of his initiative to create a coherent, national movement with which to engage the larger culture, he created *Christianity Today* in 1956. Hewing to a doctrinal middle ground, Graham said the magazine would be theologically conservative but with a liberal approach to social issues. He said, "It would combine the best in liberalism and the best in fundamentalism without compromising theologically." He condemned nonbelief as promoting lax morality and, in tune with the great threat of the times, spreading communism.[21]

Graham had maintained a milder stance on race issues than many of his fellow evangelicals in the South. But after the 1965 Watts riots in Los Angeles, he denounced the breakdown in social and moral order, distanced himself from the more liberal aspects of the evangelical movement, and became more overtly political. The riots and Graham's response to them—including aligning himself with Richard Nixon's law-and-order campaign in 1968— accelerated evangelical and fundamentalist migration into the Republican Party, especially as Northern Democrats became champions of civil rights. Graham's connections to conservative Southern fundamentalists and evangelicals merged with white voters' shift to the Republican Party in the 1960s. The shift was not only to a conservative, Christian South, but to a politically active one.[22]

Graham was only a part—but an important part—of a larger trend that saw a conservative movement realign American politics and grow into the "culture war" of the 1980s and 1990s. He was the fundamentalists' media master, a handsome, youthful man whose new fundamentalism set out to provide a philosophic, rational foundation for Christianity in an apparent effort to transcend the anti-intellectualism of older fundamentalism. He was charismatic, eloquent, and popular. He was a television natural, drawing huge audiences to his urban tent revivals. He was "new" not just in his media appeal, but also in his ability to remain committed to scriptural infallibility while drawing into the tent the opponents of political and social liberalism.[23] Graham's monumental revival in Madison Square Garden in May 1957 looked much like a political convention, organized to impress, to please the eye, to assure the masses, to reaffirm righteousness. The sensational rhetoric and personal charm were ingredients not of a mere campaign, but of a crusade. Revivalists had radio success behind them. Now, they had television. Graham added another dimension to the politics of religion as an unofficial consultant to the White House in 1952 when he advised Dwight Eisenhower to be baptized into the Presbyterian church of Eisenhower's parents. The president-elect had sought Graham's advice on tactfully handling the fact that he had never been baptized in a church. Graham became a pastor to presidents after that, giving evangelicals a new dignity and place in American government and politics.[24]

A darker perspective on organized religion also became evident in the 1950s. *Inherit the Wind*, the hit stage play (1955) by Jerome Laurence and Robert Edwin Lee and movie (1960) adapted by Nedrick Young and Harold Jacob Smith, set its anti-McCarthy message to the Scopes trial, staged with a small-minded, raving Bryan character against Spencer Tracy's heroic Darrow. Both the play and, especially, the movie portrayed fundamentalists as ignorant, backward bigots flailing against the progressive, scientific forces of science and learning. The movie came in the wake of the successful launch of the Soviet satellite Sputnik, presumably demonstrating the inferiority of American science. With the Cold War at full chill, Americans needed science—not antiscience—to triumph over godless communism.

The popular response to the movie version—the one of greater concern given exposure of stage versus movie in the broader culture—was mixed. *Time* magazine blistered the movie as a "confused manipulation of ideas and players." The reviewer found Tracy's Darrow character a mere "Holly-wooden archetype of the wise old man." Fredric March's Bryan, instead of "that unbalanced genius of the spoken word," was "a low-comedy stooge who at the climax catches a facefull of agnostic pie." The reviewer acknowledged that both Bryan and Darrow are "serious and important men in memory." In addition, he charged, "The script wildly and unjustly caricatures the fundamentalists as vicious and narrow-minded hypocrites, just as wildly and unwisely idealizes their opponents, as personified in Darrow. Actually, the fundamentalist position, even when carried to the extreme that Bryan struck when he denied that man is a mammal, is scarcely more absurd and profitless than the shallow scientism that the picture offers as a substitute for religious faith and experience."[25]

Newsweek summed up the movie as "blowing up a fine storm" with special note of the final scene that suggested the Darrow-Tracy character perhaps was not an atheist. In an interview with *Newsweek*, director Stanley Kramer acknowledged taking the atheistic edge off Darrow-Tracy, "'Cleaning up Tracy for the family trade.' . . . I read Darrow's speech at the end of the Leopold and Loeb trail and it seemed to me it wasn't made by somebody who believed in nothing."[26] However suspect the drama's historical accuracy, it resonated for posterity and became a reference point for summing up the Scopes trial, even if inaccurately.

The Genesis Flood

Fundamentalists struggled to find an effective response to *Inherit the Wind*. They may have succeeded, only a year after the movie, with a book dismissed by scientists and theologians but which sold tens of thousands to believers in its first decade. In *The Genesis Flood: The Biblical Record and Its Scientific Implications* (1961), Henry Morris and John Whitcomb blended biblical literalism and very suspect science to show the Earth was only about 6,000 to 10,000 years old, that creation occurred in a few days, and that humans had coexisted with dinosaurs. They claimed to show a great flood wiped out life on Earth except for that which escaped in the Ark. The same flood explained geologic strata, they asserted, and so "the last refuge of the case for evolution immediately vanishes away, and the record of the rocks becomes a tremendous witness . . . to the holiness and justice and power of the living God of Creation!" They called evolution bad science. Morris, chairman of the Virginia Polytechnic Institute's civil engineering department, became creationism's leading voice. He told creation-movement historian Ronald

Numbers in a 1980 interview that years earlier he began calculating the improbability of highly complex creatures after having watched the butterflies and wasps that flew in his office window. Earlier, in *That You Might Believe* (1946), Morris incorporated flood geology with young-Earth creationism in what he claimed was the first such book from a secular university.[27] It was a hit among young-Earth creationists because it included explanations of postflood fossils that were interpreted to prove human and dinosaur coexistence. Whitcomb earned his BA in history in 1948 and then enrolled in Grace Theological Seminary, a fundamentalist school in Indiana. After receiving his Bachelor of Divinity Degree, he remained at the seminary, teaching Old Testament and Hebrew and continuing graduate studies, which he finished in 1957. His dissertation was "The Genesis Flood," which drew heavily on Price's flood geology in arguing for young-Earth creationism.[28]

With *The Genesis Flood*, Morris and Whitcomb infused creationists with a new energy, fueled by the prospect of their own scientific argument via their own scientific credentials. Creationists now had their "scientific" alternative to Darwinian evolution.[29] This would be very important to the movement as it appealed to a broader constituency and would widen creationism's appeal beyond a narrow base of theological conservatives. *Genesis Flood*, creationists felt, was an answer to mainstream scientists' out-of-hand dismissal of creation-science. The Creation Research Society (CRS) started in 1963 with a group of eighteen, which included Morris. Part of the impetus was to lend scientific legitimacy to creation-science, which they attempted to do via a quarterly journal, the *Creation Research Society Quarterly*, and by limiting full membership to individuals with graduate degrees in a scientific field. People whose credentials were not up to the standards were "sustaining members."

Though not a biologist, geologist, or anthropologist, Morris was a full member by virtue of graduate training in civil engineering. More than 2,000 members had joined the society by the end of its first decade, 450 of them "regular," according to the CRS. Evolution had begun to reappear in public-school textbooks in the 1960s, and the 1963 Biological Sciences Curriculum Study (BSCS)included controversial subjects, evolution among them. The BSCS was a federally funded effort, inspired in part by the shock of the Soviet Sputnik launch, to improve the quality of high-school science textbooks. The newly created National Science Foundation, working with university scientists, found evolution largely absent from textbooks. The popularity and widespread adoption of the BSCS textbook across the country meant a revival of evolution. This, in turn, meant increasing challenges to state antievolution laws. However, antievolutionists were gaining momentum and getting the attention any campaign craves and needs.[30]

Evangelicals lost in a series of federal-court decisions from 1961 to 1971. Those rulings outlawed Bible readings and prayer in public schools, and voided state aid to religious schools. A 1968 U.S. Supreme Court decision, *Epperson v. Arkansas*, found a 1928 Arkansas statute unconstitutional because it imposed religious restrictions on teaching evolution. Susan Epperson, a tenth-grade biology teacher in Little Rock, challenged the law on grounds that it prohibited teaching human evolution. The previous year, the Arkansas Supreme Court had upheld the antievolution law in what Edward Larson called a "bizarre" opinion of only two sentences. The opinion said the law was a "valid exercise of the state's power to specify the curriculum in its public schools."[31] There was no discussion of constitutional issues.

The U.S. Supreme Court took the case by granting certiorari in 1968. The Court ruled the statute unconstitutional because, contrary to the Establishment Clause, it favored a particular religious group, one that believed evolution conflicted with the Genesis account of creation. The Court said, "The First Amendment mandates government neutrality between religion and religion, and [between] religion and nonreligion." Creationists apparently had been soundly beaten. Justice Abe Fortas, writing for the majority, said the law violated the Establishment Clause because the Arkansas legislature had a clear religious purpose in the law, and states must be neutral toward religion. But creationists showed increasing acumen in their campaign. Their solution was to demand fairness—equality, equal time, balance. However one might cast it, the plea seemed so reasonable: that censorship should not prevail, that dissent should be tolerated, and that principles of free expression should be maintained. Larson said the press hailed the ruling as a victory over the "dead hand of a bygone era."[32]

A *Chicago Tribune* story on the Epperson case reflected the range of responses, from acknowledging its constitutional significance, to being rather dismissive of the whole issue, to resurrecting Menckenesque disdain:

> Thus when a similar case reached the Supreme Court in Washington Tuesday, the court was able to raise a rather pompous cloud of dust by unanimously holding that anti-evolution statutes . . . constitute "an establishment of religion" in violation of the 1st amendment. . . .
>
> Justice Harlan remarked: "I think it deplorable that this case should have come to us with such an opaque opinion by the state's highest court. With all respect, that court's handling of the case savors of a studied effort to avoid

coming to grips with this anachronistic statute and 'pass the buck' to this court."

So the Supreme Court establishes another great landmark in the history of free inquiry by pontificating on an issue which aroused no interest in anyone—least of all, among those delegated to defend it.[33]

Scopes was still the reference point. *Time* magazine explained *Epperson* in the context of the movie version of *Inherit the Wind* and quoted 68-year-old John Scopes's approval of the Supreme Court decision: "It is what I have been working for all along." The article pointed out that no teachers, including Epperson, had been indicted under the Arkansas statute and that the ruling "should put an end to the issue in Mississippi as well."[34] A *New York Times* story included a Scopes-trial photograph of Bryan and Darrow in the courtroom. The *Times* accused the Supreme Court of attempting to dodge the issues of academic freedom raised in the case when it noted "'the multiplicity of controversies that beset our campuses today,' and it insisted that the courts can properly intervene to protect 'fundamental values of freedom of speech and inquiry and of belief.'" The *Times* and the decision reflected the '60s temper with the concern for free expression and the role of government.[35] In another story, more bland in tone, the Scopes trial ancestry of the case was duly noted, with Justice Fortas's concession that "the law had never been enforced and 'is presently more of a curiosity than a vital fact of life' in Arkansas."[36] *Newsweek*'s four-paragraph coverage of the case began with historical error—that Scopes "dared challenge the biblical version of creation in the schools of Dayton"—and ended with Scopes's endorsement of the Epperson decision. It was, in the *Newsweek* account, a local issue involving the "uncomfortable bedfellows" of biology and the Bible.[37]

Creationists were finding their own antiestablishment voice in the 1960s, demanding at the very least to be heard and, preferably, to be granted membership to what they characterized as the exclusive club of the mainstream scientific community. It was a good time to remind Americans of their egalitarianism and individualism. Creationists were attuned to the cultural zeitgeist of the decade, and their political savvy had increased substantially since Scopes. When the Epperson decision prescribed state neutrality toward religion, creationists interpreted it as meaning room for competing explanations. By then, the literalists had become the rebels, the ones offering an alternative to the establishment science. Though creationists' demand for equal time had no legal foundation, it sounded good. Dorothy Nelkin even attributed the Federal Communications Commission's Fairness Doctrine as being the origin of creationists' equal-time tactic.[38] The Fairness Doctrine did not prescribe equal time, or any particular ratio of time for different sides. The doctrine attempted to get broadcasters to help create an informed public by offering more than one side for controversial ideas of public importance. Equal time was not a new idea on the part of creationists, but even if the equal-time argument was not grounded in the Fairness Doctrine, that policy was at the very least an amplifier for smaller voices. The concern for minority rights was nothing new; Americans put the idea into practice in the 1960s in a variety of ways, perhaps most notably in the Voting Rights Act of 1964. In this respect, the creationist challenge to conventional science was well fitted to the sentiments of the decade, the challenge to government, the revolt against the establishment.

At the beginning of the 1960s, the Tennessee legislature had rejected an attempt to repeal the antievolution law under which John Scopes was prosecuted. In 1967, legislators repealed the law. Like Bryan's death at the end of the 1925 trial, the death of the law did not mean the demise of the idea.

The "Decade of Creation"

Strong currents ran in both directions. For evolution and science, a landmark 1971 case, *Lemon v. Kurtzman,* was a clear legal victory, and it established an important test for deciding whether creationist-inspired legislation was government-promoted religion. For antievolutionists, the 1970s saw the introduction of twenty-four equal-time bills in twelve states. Creationist texts were approved in six states. It was enough of a popular groundswell that Morris later deemed it the "decade of creation."[39] Creationists' tactical shift from confrontation to equal time was astute because the message resonated well, sounding more politically moderate and centrist. Creationists now had the dual forces of egalitarianism and, in their own eyes, scientific authority. In the wake of the Epperson decision, Morris told creationists to promote the scientific aspects of creationism, leaving out references to Genesis and the flood.

The campaign was shifting again. In the 1960s, creationists discovered they were antiestablishment. In the 1970s, they fitted their rebel cause to American democratic tradition. Consistent losses in court did not dampen creationist ardor. They seemed to see the legal setbacks as merely a remanding to their own court of public opinion for refinement of campaign tactics. The U.S. Supreme Court's 1971 ruling in *Lemon v. Kurtzman* erected a substantial legal barrier to promoting religion in public schools. In that Pennsylvania case, the Court delineated a three-part test, which became known as the "Lemon test," on adhering to the Establishment Clause: 1) legislative intent must be secular and cannot constrain or promote religion; 2) the law cannot constrain or promote religion; 3) the law cannot cause "excessive government entanglement with religion."[40] If the legislation failed on any single count, then it was unconstitutional. All subsequent creation-science cases have cited the Lemon case, and creationist defendants have lost on at least the first criterion—legislative intent.[41] In the press, the case often was simply one of separation of church and state. Some saw the case as more, as the *New York Times* did in its dissection of the decision, pointing to the First Amendment issues but also criticizing the opinion itself for its "divisive political potential related to lobbying for aid." Lemon concerned state aid to Catholic schools. The *Times* story referred to the "new secularism" in the context of motives for choosing public schools. Many Catholic parents had come to see the church schools as a handicap to education because they failed to reflect the larger culture. "[T]he parochial school is hardly a microcosm of the larger society. Containing neither religious nor racial mix. . . ."[42] Within a few decades, creationists twisted the concept into the argument that public schools needed to include all ideas about origins that local parents deemed appropriate. The *Los Angeles Times* quoted the opinion's concern about the divisiveness such laws would engender, inviting "political division along religious lines. It would put politics into religion and religion into politics." *Time* magazine also considered the politics of the case, the opposition of popular opinion, and Billy Graham's support for public aid to church schools to "counterbalance the 'materialistic, atheistic teaching' in public ones."[43]

The *Lemon* decision did not dissuade creationists from their mission, but it did push their campaign in a different direction, which was evident in a piece of Tennessee legislation. A 1973 equal-time law did not outlaw teaching evolution, as had the law under which Scopes was convicted, but it did require public-school textbooks to present theories of human origins as "not represented as a scientific fact" and to provide equal space for alternative theories, which included Genesis. The author of a creationist textbook, Russell Artist, a biology professor at Churches of Christ–affiliated David Lipscomb College in Nashville, initiated the action with an argument that sounded eminently logical and rational, and which appealed to fairness: "We're not telling them not to teach it [evolution]—let them teach it all they want. . . . Now is that being prejudiced? No, it's simply asking that we should have the same amount of time to present our creation point of view."[44] Joseph Daniel was one of three biology teachers, along with the National Association of

Biology Teachers, who filed suit in federal court shortly after the bill became law. A parallel action arose in state court with a challenge by Americans United for the Separation of Church and State. The resulting legal tangle was straightened out quickly when in 1975 the federal court of appeals declared the law unconstitutional because it was preferential for the biblical version of creation. The court heard no arguments in the case. It was another loss for creationists, but many of them saw the ruling as pointing out some technical issues—such as reference to Genesis, excluding satanic ideas—that could be fixed, clearing the way for new equal-time legislation.[45]

One of the central players in the creationist revival a few decades earlier claimed to have been rather apolitical, not advocating legal or legislative action. Henry Morris said his Institute for Creation Research did not initially seek to aid local activists, but did not deny the help when they came for it. According to Witham, Morris later said he had concerns about attempting to compel people, under penalty of law, to hear the creationist side.[46]

By the 1980s, creationists were political, in spite of consistent courtroom losses and public/media skepticism. The most prominent political-religion movement was Falwell's Moral Majority, created in 1979 by political conservatives who were frustrated with the liberal drift of the Republican Party. They wanted to confront the liberalism that came into American society during the 1960s. Falwell was an ideal representative—given his large congregation, huge television audience, and charisma. The Moral Majority was not a strictly fundamentalist organization because the political goals—building a more conservative majority—meant building a coalition among many groups, including Catholics, Jews, Mormons, and even secularists. Practical politics drove the coalition, because at the time evangelical Protestants made up only 15 to 20 percent of the population. Such pluralism cost some support among more hard-line fundamentalists, but the message had a broad appeal: an "immoral minority" had displaced the Bible as the foundation of American society. A number of objections defined the coalition: antiabortion, antidrugs, anti–gay rights, anti-détente with Soviets, anti–Equal Rights Amendment. The Moral Majority made sure people registered to vote and worked to influence that vote. But more than just conducting its own campaign, the Moral Majority was teaching its members to deal with media and to become involved in public life, which included running for office. Their results were mixed. Most notably, and thanks in no small part to the Moral Majority's efforts, the Equal Rights Amendment failed to pass. Several states that were strongly lobbied by the Christian Right reversed their previous endorsements: Tennessee, Kentucky, Indiana, South Dakota, and Nebraska. Another prominent objection among those in the movement was teaching evolution in public schools. Conservative Christians helped persuade legislatures in Arkansas and Louisiana to pass bills that gave equal time to Genesis and evolution.[47]

The Emergence of "Intelligent Design"

Two particularly important aspects of what was to become twenty-first-century creationism first emerged in the 1980s: Creationism became part of a larger, well-defined, well-organized political movement, as seen in its inclusion in the Moral Majority agenda. Secondly, "creation-science" evolved into "intelligent design." Creationists believed intelligent design was the way to get around the Establishment Clause and objections to teaching religion in schools. Intelligent design also was a good public platform for promoting a "scientific" alternative to evolution. The target for such a message was not mainstream science but the larger public, often only semiliterate in things scientific. The *Bible-Science Newsletter* in March 1980 advised its creationist advocates to emphasize science: "Sell more SCIENCE. . . . Who can object to teaching more science? . . . do not use the word 'creation.' Speak only of science. Explain that withholding information contradicting evolution amounts to 'censorship.' . . . Use the 'censorship' label as one who is against censoring science. YOU are for science."[48]

The 1980s started out very well for creationists. The soon-to-be-president Ronald Reagan believed evolution and creationism should share time in public schools. Numerous surveys showed large numbers of people believing in creationism, though the survey results often were distorted by poor question wording, or ill-defined terms, such as "creationism" itself.[49] A 1982 Gallup Poll showed 44 percent of Americans believing "God created man in present form."[50] The decade also began with a decisive defeat for creationism, in which a U.S. District Court ruled in *McClean v. Arkansas Board of Education* that creation-science was not science. The decision set back the argument for equal-time pleas when judges rejected the law titled "Balanced Treatment for Creation-Science and Evolution-Science" Act. In a more nuanced argument for equal time, antievolutionists exploited the idea of strict neutrality by the state in religious issues. They argued that banning creationism from science classes was, in effect, not neutral but favoring one side—the evolutionists. Thus, the creationist argument went, alternative views should be permitted in the interest of fairness and freedom of expression, In the 1982 decision, U.S. District Court Judge William Overton ruled the law was intended to advance religion with its reference to God and supernatural creation. He cited the lack of peer-reviewed, published research, the fact that creationism did not consider opposing data or research, and the fact that its foundation was the literal word of the Bible. Overton said Act 590, signed into law by the Arkansas governor in 1981, was unconstitutional because it was passed with the purpose of advancing religion, it promoted "particular religious beliefs, and it entangled the state of Arkansas with religion."[51]

As had become standard media practice in cases involving public schools and antievolutionism, the press tagged *McLean* "Scopes 2." Historian Marcel LaFollette found the press treated the trial simply as a Scopes replay, to the point of naming John Scopes more than any witness in the McLean trial itself.[52] As though to justify the "Scopes 2" label, photos of Darrow and Bryan appeared regularly in the trial coverage. Creationists won their campaign in one respect when the *Arkansas Democrat* criticized the judge for allowing the ACLU to bring religion into the trial, thereby impeding the "free flow of ideas." The *Democrat* also warmed to creationism's "underdog" role, the beleaguered minority set upon by unyielding mainstream scientists. The ACLU, the newspaper noted, was not championing the underdog in this case but was in lockstep with the status quo and convention.

The *Arkansas Gazette* took the opposing view, reporting on the nonscientific aspects of creation-science, which it viewed as religion. The newspaper commended the ruling for reaffirming the separation of church and state.[53] The creation campaign succeeded because it won public and media attention and some support. And it won with the themes it had introduced decades earlier, themes that resonated in the broader culture, no matter the opinions of "conventional" scientists. The media had begun to treat the trial not as a science issue but as a political issue. Most major newspapers covering the trial did not quote scientists who were called to testify about the veracity of creation-science—a lineup that included Stephen Gould. Only the *Boston Globe* and the *Washington Post* carried articles by science writers. Instead, the assignment often fell to a political or legal affairs reporter. The social and political aspects of the trial then became the focus of much of the coverage, in spite of the fact that so much of the state's case depended on showing creation-science to be science.

A subsequent lawsuit showed creationists' perseverance and their growing political sophistication as they tested the Arkansas ruling, looking for openings it may, or may not, have provided. In *Aguillard v. Treen*, a 1981 Louisiana lawsuit stopped an attempt to include creation-science as a "balanced treatment" in public schools. Attempting to avoid the losing argument in the Arkansas case, Louisiana's new law left the idea of origins open, without reference to biblical stories, and delegated origins to an earlier time "by complex initial appearance" and allowed "scientific creationism" to be balanced with evolutionary explanations. The Louisiana law did not say what "creation" meant, nor did it mention God. The case meandered through state and federal courts for five years before resolution.[54]

In July 1985, the federal district court, in a summary judgment without public trial, found the legislation unconstitutional because the state, again in violation of the Establishment Clause, failed to be neutral on religion. The court found creationism was a religious doctrine. On appeal, by which time the case had become *Edwards v. Aguillard* by virtue of a new governor having been elected and becoming the new plaintiff, the U.S. Court of Appeals for the 5th Circuit upheld the decision. In June 1987 the Supreme Court upheld (7–2) the summary judgment, finding the Louisiana law unconstitutional because the legislature passed it with the intent of having Genesis taught in public schools. The ACLU proclaimed it the "death knell" for creation-science. It was not.[55]

The defendants' essential argument was that, unlike the Arkansas law mandating teaching of creation-science, the Louisiana law was not explicitly religious. Even if on the losing side, though, the defense was fully cognizant of the benefits of publicity—pro or con. Though suffering a major legal defeat, creationists had gained political and public traction. The benefit of getting into a losing fight with mainstream scientists was credibility. Simply having a place on the stage endowed one with some degree of legitimacy. It was akin to a third-party candidate being part of a nationally televised presidential-candidate debate, as was the case with Ross Perot in 1992, when he had no chance of winning. But he did win audiences and converts. His ideas, in particular about federal deficits, seeped into the national discourse and remained there, requiring responses from the electable candidates. Similarly, creationists were losing cases but had become part of the national agenda.

In 1987, at the outset of the *Aguillard* case, the *New York Times* immediately resurrected Scopes—and not just "Scopes 2." Instead, it published a page on religion and science with columns drawn from Clarence Darrow's remarks on the second day of the trial and H. L. Mencken's jottings for the *Baltimore Sun* under the heading "Scopes: Infidel." Above both, Stephen Gould critiqued "scientific creationism" as merely biblical literalism and recounted his own testimony in the Arkansas case to that effect.[56] Finally, years later, covering the U.S Supreme Court in *Aguillard*, the *Times* reported the severity of tone in Justice William Brennan's dismissal of the equal-time rationale as a "sham." But, the story balanced that side of the argument—the majority of the court—with coverage of the dissent that denounced the majority decision as "repressive" and "illiberal," "preventing the people of Louisiana from having 'whatever scientific evidence there may be against evolution presented in their schools.'" The dissenting opinion also alluded to the minority opinion's concern about attributing improper motives to "'the democratically elected representatives of the people.'" Here, apparently, the creationist plea for equal time and democratic process had found sympathy. The *Times* and other newspapers consistently referred to the Scopes trial in reporting the *Aguillard* decision but had come to understand the case as a political one, where a science-religion debate was part of a larger political landscape, more than an Establishment-Clause question. The issue reflected "considerable political and judicial activity," according to the *Times*.[57] The *Washington Post* also reported the tough words for creationism in the majority opinion but gave space to the losing side. It quoted lead attorney Wendell Bird for the creationists. He said the majority opinion did not rule out teaching creation-science: "'The justices did not say it was inherently unconstitutional to teach creation science along with evolution, just that the particular purpose of the Louisiana legislature' was unconstitutional." The *Post*, too, was taken by the appeal to democratic values, going on to quote the attorney: "'Eighty-six percent of the public favors "balanced treatment" laws,' and '86 percent of the public can't be stopped in the long run. . . . You can't stop anything so broadly supported, only delay it,' he said."[58]

The court cases helped accelerate the evolution of "creation-science" into "intelligent design," which emerged in the mid-1980s. Creationists promoted it as a scientifically valid alternative to materialistic evolution. Intelligent design's proponents adopted scientific language in promoting divine creation, which they saw as evident in the existence of complex organisms that functioned only as a whole and were worthless without all the intricacies. It was an old idea that predated Darwin, and the new books were updates of William Paley's early-eighteenth-century watchmaker analogy. His argument was that a design implies a designer. Paley wrote that if one came across a watch on the heath, then it was logical to assume someone dropped it, and a watchmaker had made it. The watch was too complex to have arisen by natural processes. Two publications, in particular, of the mid-1980s brought intelligent design into the fight. The first of the modern incarnations of the watchmaker analogy and the introduction of intelligent-design terminology was *The Mystery of Life's Origins* (1984), which declared that natural causes could not explain life's origin, emphasized problems with evolutionary theory, and did so without reference to God, the Bible, or Genesis. There was only a mention in the Epilogue of "intelligence" being involved in life's origins.

The second book was *Of Pandas and People*, which appeared in 1989.[59] Its authors originally saw it as a scientific defense for creationism, but revised their manuscript in light of the Supreme Court's *Aguillard* ruling. The authors substituted "intelligent design" and "design proponents" for "creation" and "creationists." The book title also had changed, from *Biology and Creation*.[60] Favorite examples of intelligent design's existence were the bacterial flagellum or the human eye—neither of which would be functional, proponents argued, without all the parts in place and functioning, i.e., like a watch. It was an appealing (but false) argument because it simplified the complex and provided solace to those discomforted by the possibility of evolution's truth. It would be easy to sell in a mass-mediated culture.

As though to complete the recasting for the old drama, Richard Dawkins published *The Blind Watchmaker* (1986), which made the empirical case for complexity without God. It was the intelligent-designers for the literalists and Bryan, Dawkins for science, evolution, and Darrow. Dawkins was not alone in answering creationism. He stood out in popular appeal and severity of tone. Anointed "Darwin's Rottweiler" by the man who endowed Dawkins's professorship at Oxford University, Dawkins has conducted a strident frontal assault on creationists, whom he has called "ignorant, stupid or insane." He compared faith to smallpox—"but harder to eradicate." Genesis? Just another myth "adopted by one particular tribe of Middle Eastern herders."[61] His books have been reviewed favorably and read widely, which may be more a testament to an American audience's willingness to accommodate instead of any sea change of ideas.

A Political Religion

Fundamentalists were well attuned to the political times in the resurgent conservatism of the 1980s. Many had, as in the 1930s, repackaged themselves as evangelicals, reaching out to a broader constituency. Their conservative theology often meant biblical literalism, and was part of a package of issues, including abortion, home schooling, and smaller government. The court challenges continued, but creationists turned increasingly to politics to promote religion in schools. The issues under fire in the new "culture war" shared several overarching themes, including individual rights and the sanctity of life. Neither is a scientific issue, but both are impinged upon by advances in science and technology, and by changing attitudes toward traditions. In several respects, it was the same concerns of antimodernists in the 1920s, when many saw their traditions and beliefs discredited, dismissed, or condemned.

Though still only a replay of the Scopes trial, the creationist movement was more politically sophisticated than the antievolution/fundamentalists of the 1920s. By the late 1980s, creationists were no longer merely opposed to evolution. Instead they were "pro" science, recasting themselves as a democratic exercise, and with God's blessing. What they lacked in intellectual rigor they made up for with popular appeal. Besides, intellectual integrity was not the issue. Unlike the 1920s fundamentalists, creationists of the 1980s and 1990s promoted themselves as the real defenders of America's frontier spirit, venturing into scientific realms shunned by the hidebound, timid mainstream scientists, who were assailed as being locked blindly into conformity. The Moral Majority also provided a political engine with wheels in place of the creationists' intellectual oxcart. Creationists were no longer an isolated, single-issue fringe group, but part of a larger movement tied to larger Republican political goals and agendas.

According to Gallup Polls since the 1970s, anywhere from 40 to 50 percent of Americans have remained steadily and solidly skeptical of evolution and embraced young-Earth creationism. In the Gallup surveys from 1982–2008, 43 to 47 percent of respondents agreed with the statement "God created human beings pretty much in their present form at one time within the last 10,000 years or so." By comparison, 35 to 40 percent selected the statement, "Human beings have developed over millions of years from less advanced forms of life, but God guided this process." Nine to 14 percent selected the option that humans had developed over millions of years, "but God had no part in this process."[62]

Bryan may have stumbled at the Scopes trial, but he had been a successful, substantial politician in his day. Fundamentalists grasped his lesson—appeal politically to cultural values, win headlines, and be more than a naysayer. Bryan won when he appealed to the masses, not when he fought intellectuals. Fundamentalists appeared to have fathomed this tactic when their audience became a broad political constituency rather than judges and scientists. Courtroom losses and scientific condemnation became rallying points for individualism, freedom of expression, and antielitism.

4

Intelligent Design and Resurgent Creationism

It is absolutely safe to say that if you meet somebody who claims not to believe in evolution that person is ignorant, stupid or insane (or wicked, but I'd rather not consider that).

—Richard Dawkins

We need to replace Dawkins-style and Sagan-style science with a science that is humble about what it can do.

—Phillip Johnson

Defining "creationism" had become even more problematic by the 1990s. The term could encompass the traditional "young-Earthers," who believed life and Earth were created spontaneously less than 10,000 years ago and that humans came into being in their present form. The term also could include those who accepted an old Earth, thereby accommodating geologic facts. Political expediency, though, favored a broader definition in order to win a larger constituency. Up to this point, politics had lagged behind religion as a motivating force.[1] That, too, was changing, as creationists were turning to more overtly political goals, which could be seen in the New Religious Right of the 1970s and the Moral Majority of the 1980s. Public opinion polls showed that nearly half of Americans believed some variant of creationism to be a viable scientific alternative to evolution.[2] Creationists made individual rights and freedom of expression consistent themes in the fight to teach creationism in public schools. This put scientists and their allies in the awkward position of telling a culture based on individual liberty and civil rights that such values were irrelevant to this debate.

Antievolutionism was thriving in spite of losing decisively and consistently in court over several decades. Creationists were learning from the legal setbacks. The most important lesson: It was not a battle for scientific respect, which they kept losing, but a political campaign. Creationists began treating it as such. In 1984, the National Academy of Sciences sent to science teachers and school officials nationwide more than 40,000 copies of a booklet, "Science and Creationism: A View from the National Academy of Sciences." The Creation Research Society (CRS), in turn, urged the American Scientific Affiliation (ASA) to issue its own pamphlet: "Teaching Science in a Climate of Controversy: A View from the American Scientific Affiliation." To anyone unfamiliar with the pedigrees of the groups, it must have appeared to be a debate between two groups of scientists. But the ASA was largely an antievolution, evangelical group. Their pamphlet sounded moderate as it urged teachers to explore the "broad middle ground" where God and science could coexist.[3]

Creationists moved away from arguing with scientists and toward providing a scientific-sounding argument for mass consumption. They did not abandon the assertion that science must be subservient to biblical truth. Instead, they cloaked the dictate in the language of science, usually in the guise of "intelligent design." The term apparently first emerged in 1984 in the *Mystery of Life's Origins*.[4] The chapter titles included "Crisis in the Chemistry of Origins" (Chapter 1), "Simulation of Prebiotic Monomer Synthesis" (Chapter 3), "The Myth of the Prebiotic Soup" (Chapter 4), and "Thermodynamics and the Origin of Life" (Chapter 8). The authors mentioned the possibility of supernatural origins only in the Foreword and Epilogue. In the Foreword, Dean Kenyon outlined the problems that have dogged researchers in origin-of-life research, especially in chemistry laboratories. Kenyon, a biology professor at San Francisco State University, in 1989 coauthored *Of Pandas and People* with Percival William Davis. It was the first biology textbook to advocate intelligent design. In *Mystery*, he cited the "enormous gap between the most complex 'protocell' model systems produced in the laboratory and the simplest living cell." "[T]here is a fundamental flaw in all current theories of the chemical origins of life." That flaw was the failure to consider the supernatural. Kenyon praised *Mystery* for its critique of the field and for addressing problems, particularly the spontaneous origin of life "by purely chemical and physical means."[5] The Epilogue summarized the alternative views of life's origins as

1. new natural laws
2. panspermia
3. directed panspermia

4. special creation by a creator within the cosmos
5. special creation by a creator beyond the cosmos

The first, which included evolution, was dismissed as a failure. The second was the "extraterrestrial view . . . a life spore was driven to earth from somewhere else in the cosmos by electromagnetic radiation pressure." This, too, was dismissed because the theory simply pushed the problem of origins to another place in the cosmos. The third explanation was similar to the second, but added spaceships and extraterrestrial beings, making Earth into a sort of intergalactic dumping ground. However, the author noted, it also just pushed the problem of origins to another place. Not surprisingly, alternatives four and five were given more credence, but it was noted that whether or not "an intelligent Creator did create life . . . is beyond the power of science to answer." The authors pointed out, "Another question which can be answered . . . is whether such a view as Special Creation is plausible." It was, they concluded. The book ended with a plea for tolerance, admitted the need for more research, and declared the limits of naturalistic science.[6] It was not an empirically grounded defense, based on scientific method, but a political-cultural one that appealed to inclusiveness, tolerance, and humility.

Nonscience for the Masses

The end of the twentieth century witnessed the emergence of a body of popular books advocating creation-science. These books were at the center of a push to become scientifically respectable—not via the scientific establishment's conventional channels of peer-reviewed research in journals and books but by appeal to the scientifically unlettered. It was for the masses, and it mattered not that creation-science/intelligent design had no scientific credibility. The endorsement of enough people meant, for the idea's advocates, that it was legitimate. The tactic was a rejuvenated Bryan-Jeffersonianism. Bryan's radical interpretation of the will of the people meant science, like everything else in society, was subject to the will of the people.

Reflecting court decisions and political reality, the leading creationist textbook, *Of Pandas and People* (1989), substituted *intelligent design* for *creation* and omitted such issues as age of the Earth and fossil evidence. The book's high point (or low point, depending on one's perspective) was its place at the center of the Dover, Pennsylvania, trial, in which a federal judge found intelligent design was not science, but religion. Kenyon and Williams distinguished themselves from some parts of the creationist movement by stating in *Pandas* that intelligent design was not just revised fundamentalism. Intelligent-design proponents sidestepped arguments about interpreting scripture by simply not mentioning God or Genesis. Such tactics did not mollify critics, who accused intelligent-design advocates of "stealth creationism." But there were differences among creationists. Most significantly, creation scientists insisted on a recent creation of less than 10,000 years and a Noachian flood. By contrast, many intelligent-design proponents could accommodate a much older Earth and a less literal reading of scripture. It may have been a distinction without a discernible difference where the general public was concerned. Even the *New York Times*, which was well attuned to the creationism controversy, did not distinguish between creationism and intelligent design. Reporting on the Tennessee legislature's consideration of a bill in 1996 that would allow boards to dismiss teachers who teach evolution as a fact rather than a theory of human origins, the *Times* wrote, "Proponents of what is usually called either 'creation science' or 'intelligent design' say there are so many anomalies and mysteries of the origin of the universe and the development of life that theories other than evolution must be considered."[7] Many creation-scientists did not like intelligent design because they felt it might marginalize the Bible. But the shift from creationism to intelligent design meant antievolutionists would appeal to a broader audience.[8] The apparent elasticity of creationism was important in winning a wider constituency, particularly if the idea was offered as an alternative to what creationists deemed a rigid, materialistic philosophy—i.e., evolution—that precluded God. It was not important that creationists be scientific. It was important that they sound scientific, for the same reason that Bryan needed to embrace science. It was good politics to sound progressive, and one measure of progress in American culture has been science and technology.

Foundation for a New Challenge

Two books in particular reenergized the fight against evolution and became central to the new offensive: Phillip E. Johnson's *Darwin on Trial* (1991) and Michael Behe's *Darwin's Black Box* (1996). Johnson took evolutionists to task for what he claimed was flawed logic. Then a law professor at the University of California, Berkeley's Boalt Hall, Johnson approached the problem in lawyerly fashion. He argued that Darwinism limited the range of possible explanations by a priori excluding any metaphysical ones. Johnson focused on gaps in the fossil record and on the origins of life. *Darwin on Trial* attracted a lot of attention but offered nothing to replace Darwinism. Scientists scoffed.

The book sold hundreds of thousands. The first footnote to the volume illustrated the political dexterity of the movement:

> In this book, "creation-science" refers to young-earth, six-day special creation.
> . . . Persons who believe that the earth is billions of years old, and that simple
> forms of life evolved gradually to become more complex forms including
> humans, are "creationists" if they believe that a supernatural Creator not only
> initiated this process but in some meaningful sense *controls* it in furtherance
> of a purpose. As we shall see, "evolution" (in contemporary scientific usage)
> excludes not just creation-science but creationism in the broad sense. By
> "Darwinism" I mean fully naturalistic evolution, involving chance mechanisms guided by natural selection [emphasis in original].[9]

The note was commendable as political strategy because it embraced everyone and anyone who might accept God, and excluded only that small minority—in American culture—that is purely materialistic. He welcomed young-Earth and old-Earth creationists, brought into the fold strict literalists as well as those who might accept evolution for all of life except themselves and their progenitors, and even those who did not care or were willing to pass the problem off to God.

Johnson said the Scopes trial was the origin of the modern controversy. He called it the "great landmark" in the twentieth-century Bible-science conflict. Most Americans knew the case, he wrote, via *Inherit the Wind*, replete with religious fanatics, a heroic attorney fighting superstition in the midst of a media circus. The historical reality, Johnson correctly surmised, was a moderate Bryan, "not an uncompromising literalist," but a man who opposed Darwinism because it sanctioned German militarism and ruthless capitalism. Admittedly, the "agnostic lecturer Clarence Darrow" humiliated Bryan on the stand, which was the main purpose, according to Johnson. The film version successfully conveyed the unjust persecution of an "inoffensive science teacher" and the heroism of Darrow, "who symbolized reason itself in its endless battle against superstition." Johnson wrote that the Tennessee antievolution legislation and laws like it around the country were not a threat, but only "symbolic" measures. He believed the trial was a public relations triumph for evolution, largely because of Mencken's sarcastic coverage for the *Baltimore Sun*.[10] Generally, Johnson was right about the misrepresentation of the historical facts in *Inherit the Wind*. He was correct about Bryan, and probably gave Mencken too much credit for swaying public opinion. But his précis generally was accurate concerning the major figures and their places in the drama that became the reference point in the national debate.

In less than a decade, Johnson's devotees recognized him as a central figure in the creationism movement and decreed *Darwin on Trial* the cause's seminal book. In a 2004 testimonial dinner at Biola University, speakers gave witness to Johnson's significance in rejuvenating the movement. *Darwin's Nemesis: Phillip Johnson and the Intelligent Design Movement*, which grew out of presentations at the dinner, deemed him a "fearless leader" and "farseeing visionary . . . the fledging movement's field marshal." Fitting Johnson to a tradition in national mythology, William Dembski, in the Preface, said Johnson was an individualist "without pretense" who was taking on academe's ossified establishment. Dembski, philosophy professor at Southwestern Baptist Theological Seminary and a senior fellow at the Discovery Institute, praised Johnson for fighting modernism and avoiding "the postmodern mistake of confusing just any moving story with reality." Another contributor said Johnson was a leader of the "underdogs," who were under attack by those in "positions of power and authority." Johnson believed the materialists were operating from "fortified positions" in public schools and in the public arena. But, "The unwashed masses are not with them." "Phil saw himself appropriately as the intellectual architect of this movement." He was more than intellectual inspiration, though, as he organized people via retreats, conferences, and internet listservs; found financial support; and engineered publicity. He did so in the spirit of "eschewing authority," a "master strategist" in the fight to unseat Darwinism.[11] The alignment against modernism and with the rebels and underdogs, against the establishment and with the democratic will, was reminiscent of Bryan, not just in ideals but in elevating Johnson to the movement's leadership.

Michael Behe provided the movement its second major catalyst of the decade with *Darwin's Black Box*, in which he introduced "irreducible complexity." Behe drew an analogy to a mousetrap, which would be useless without all its parts, and which had a designer. This meant that certain organs, such as the eye, would not function if any part were taken away. Thus, it could not have evolved because without all the parts functioning at once, it had no survival value. Biologists disagreed, and noted that even a light-sensitive cell could impart an advantage over an organism lacking any sensitivity to light. Nevertheless, irreducible complexity gave creationists a scientific-sounding alternative to Darwinian evolution.[12] A Lehigh University biochemist, Behe worked from the cellular level, and he claimed to make his case based on his own scientific observations, not religious beliefs. *Christianity Today* awarded *Darwin's Black Box* its "Book of the Year Award," as supporters christened Behe a modern-day William Paley. The comparison had merit, because Behe's idea was an argument from design, akin to Paley's watchmaker analogy. Behe allowed for the possibility of an old universe, but asserted that Darwin could not explain complex, biochemical machinery at the cellular level. Behe was readable and even entertaining in his argument for intelligent design:

> Imagine a room in which a body lies crushed, flat as a pancake. A dozen detectives crawl around, examining the floor with magnifying glasses for any clue to the identity of the perpetrator. In the middle of the room, next to the body, stands a large, gray elephant. The detectives carefully avoid bumping into the pachyderm's legs as they crawl, and never even glance at it. . . .
>
> There is an elephant in the roomful of scientists who are tying to explain the development of life. The elephant is labeled "intelligent design."[13]

In spite of the argument from false analogy, it was engaging and, more importantly, accessible to nonscientists. And it must have been of some solace to anyone discomfited by the complexities and implications of science. The breezy logic was far friendlier and much less menacing than an adversary whom Behe anticipated and answered in advance: Richard Dawkins, professor of public understanding of science at Oxford University (1995–2008). Dawkins, Behe said, was the leading popularizer of Darwin. Among other things, Behe stated, Dawkins failed to explain the evolution of the human eye. Behe appeared exasperated with Dawkins and referred to Dawkin's remark that Darwin made it possible to be an "intellectually fulfilled atheist." "The failure of Darwin's theory on the molecular level may cause him to feel less fulfilled, but no one should try to stop him from continuing his search."[14] Like Johnson, Behe found himself to be the tolerant one. Daniel Dennett, materialist philosopher and author of *Darwin's Dangerous Idea* (1995), also drew Behe's ire. Both Dawkins and Dennett had opened themselves up for accusations of intolerance with their own intemperate remarks about believers, and Behe took advantage, to no gain in his scientific argument but to advantage in his political debate:

> [Intolerance] comes about only when I think that, because I have found [truth], everyone else should agree with me. Richard Dawkins has written that anyone who denies evolution is either "ignorant, stupid or insane (or wicked—but I'd rather not consider that)." It isn't a big step from calling someone wicked to taking forceful measures to put an end to their wickedness. . . . Dennett compares religious believers—90 percent of the population—to wild animals who may have to be caged.[15]

He accurately quoted Dawkins, but inflated the implications of the remark to include force against religious people. Though abrasive and daunting in his eloquence, Dawkins never advocated such things. The reference to Dennett was extremely misleading. Dennett, in *Darwin's Dangerous Idea*, did compare the necessity of caging a dangerous animal in a zoo, such as a lion, to that of caging potentially dangerous ideas. But he said more:

> I love the King James Version of the Bible. My own spirit recoils from a God Who is He or She in the same way my heart sinks when I see a lion pacing neurotically back and forth in a small zoo cage. I know, I know, the lion is beautiful but dangerous; if you let the lion roam free, it would kill me; safety demands that it be put in a cage. Safety demands that religions be put in cages, too—when absolutely necessary. We just can't have forced female circumcision, and the second-class status of women in Roman Catholicism and Mormonism, to say nothing of their status in Islam. . . .
>
> . . . We preach freedom of religion, but only so far. If your religion advocates slavery, or mutilation of women, or infanticide, or puts a price on Salman Rushdie's head because he has insulted it, then your religion has a feature that cannot be respected. It endangers us all.[16]

Behe ignored much of Dennett's argument, which was far more complex than Behe admitted.

Creationists looking for scientific legitimacy embraced irreducible complexity, while scientists criticized Behe for offering up rebranded creationism. Dawkins lambasted him for laziness for resorting to intelligent design rather than looking for scientific causes.[17] As with Johnson, Behe's appeal for a popular audience was not just his defense of religion, but that he was easy to comprehend: that a structure such as the eye could not be made simpler and still be functional. Take away any part of it, and it would be useless. Therefore, it must have been designed in all of its complexity, with multiple parts. Design implied a designer.

Just an Alternative Theory

Creationists had struggled in the past with an image problem, which began with Mencken's merciless and enduring portrayal of rural Tennesseans as hillbillies and rubes. By the 1990s, however, mainstream media were depicting creationism as simply the other side in the evolution controversy. *Darwin on Trial* was part of the turnaround—written by a professor at a major university and getting attention in the mainstream press, which was largely critical of the book. It appeared to matter little to the general public that mainstream scientists ignored or condemned the work, because it no longer was a science issue. It was a public debate about a policy, which was teaching creationism in public schools. Creationists had a publicity triumph because there was no controversy among scientists. In addition, gaining space in the press meant winning some degree of legitimacy, akin to a political opponent being a legitimate voice of opposition. Creationists had won attention and, however undeserved, scientific credibility with an idea that sounded like science, even if it was not. Scientists, science writers, and media were disdainful. The headlines continued, and the publicity swelled.

Johnson and Behe were important in revitalizing the movement because they brought credentials to the debate, were articulate, and their theses were comprehensible for the general public. They effectively presented themselves as liberal and tolerant, in contrast to several new public faces of evolution, including Dawkins and Dennett, whose atheism marginalized them for many people. But Johnson and Behe wrote for the broader culture, an audience without the time or inclination to immerse themselves in intellectual complexities, let alone ideas such as materialism that might disrupt the solace provided by intelligent design. In addition, Johnson and Behe were appealing to a culture that thrived on its history of rebels and rebellion. Johnson and Behe became the contrarians, fighting staid scientific convention. It was a tactic reminiscent of Bryan. Critical news coverage meant only that they were more visible to more people and that they were taken seriously by mainstream media covering serious science and major political issues.[18]

The press coverage tended to be rather uncritical, presenting Johnson and Behe as simply another voice in a debate. The *Houston Chronicle* called Johnson's work a "scientific challenge to Darwinism." "He says the religious community is best served by studying evolution and learning scientific concepts. Objective study will show the theory's weaknesses and flaws, which must be challenged publicly, he says."[19] Terry Mattingly, religion writer for Scripps Howard News Service, cited Johnson in a story about John Paul II's pronouncement in 1996 of the Catholic church's acceptance of evolution. Johnson said the pope did not endorse evolution and was not an ally of evolutionists. Mattingly seemed sympathetic to creationist assertions that evolution itself was a religion: "The Darwinist establishment uses arguments that betray a metaphysical or eternal point of view. From this it is a short leap to a big question: Are Darwinists using the public-school pulpit in ways that violate constitutional prohibitions against governmental advocacy of religion?"[20] The *Boston Globe* gave *Darwin on Trial* credit for being both sophisticated and politically influential. The paper quoted Eugenie Scott, director of the National Center for Science Education (NCSE), as saying several political victories had given the religious right a new base of power.

"And anti-evolutionists have published sophisticated attacks on evolution, including a book that has sold well, 'Darwin on Trial.'"[21] Such coverage simply set differing points of view against one another in a public arena.

There was criticism. A *Harper's Magazine* article on creationism, "On earth as it is in heaven," called Johnson's work "scientific could-bes: evolution has problems explaining, say, the absence of transition species in the fossil record, so evolution might be wrong and there could be a designer."[22] The *Washington Post* wrote in 1997 there was no controversy among scientists about evolution, and Behe's argument was wrong. "Behe essentially contends that, if you can't imagine how something could have happened naturally, then that is proof that the thing must have happened supernaturally. In science, ignorance is no more evidence than was Darwin's astonishment about the eye."[23]

The *Dallas Morning News*, writing in advance of Johnson speaking in the city, said he and Behe had "taken the battle further into the mainstream in recent years."[24] In this respect, Behe and Johnson became "mainstream" by implication, and being mainstream politically could eventually mean mainstream in other respects, perhaps even science.

In the *Atlanta Journal-Constitution*, Behe sounded moderate, intellectual, sophisticated. A story about a forum for Christian faculty at the University of Georgia and elsewhere said group activities included sponsoring campus speakers. "In February, they're bringing Lehigh University biochemist Michael Behe to discuss his controversial book, 'Darwin's Black Box,' described in the journal *Nature* as 'a new and more sophisticated version of scientific creationism.'" A member of the group said, "We want people to know that you can come into a university and can be fully engaged spiritually and intellectually at the same time.'"[25]

The *New York Times* gave creationists some latitude when it reported the new breed of creationists were Christian intellectuals, and did "not dispute that the planet is ancient." "But they are promoting the idea that living organisms and the universe are so impossibly complex that the only plausible conclusion is that an omniscient creator designed it all on purpose." The article cited Johnson and Behe as among the group of "energetic evangelical academics who have long been resentful that American academia gives religion no respect." Johnson said he was challenging not just university secularism, but "an entire culture that he says rests on the scientific assumption of 'naturalism.'"[26]

Johnson and Behe also took advantage of television appearances to promote the scientific credibility of creationism/intelligent design in several ways. First, they were guests on programs that usually dealt with genuine science. Second, the other debaters or panelists were sometimes prominent scientists or intellectuals. Finally, the nature of the media platform itself gave them an opportunity to sound scientific without degenerating into logic or scientific method. In a 1997 PBS *Firing Line* debate, Johnson and Behe were part of a creationism debate panel that included William F. Buckley, conservatism's intellectual beacon for a generation of Americans and founder of the *National Review*. The opposition panel of proevolutionists included Eugenie Scott; Michael Ruse, philosopher and author of several books on Darwin and evolution; and Kenneth Miller, biologist and author. Before the debate even commenced, Johnson and Behe had prevailed because the prestige of the program and other guests afforded the two with some degree of credibility and because such a lineup on *Firing Line* endowed intelligent design with the trappings of substance. A debate format puts everyone on equal footing and presumes the expertise of all involved. Johnson and Behe sounded scientific in the debates. Johnson charged that many scientists considered intelligent design unscientific not because of the lack of evidence, but because mainstream science clung to a philosophy that "excludes that Designer from reality." Scott pointed out to him that the nature of science precluded talk of "who did it" and "ultimate cause." She even agreed that Dawkins was going a bit too far with his advocacy of atheism because he mixed his philosophical views with science. But Johnson pressed the issue that "irreducible complexity" necessitated a designer, not a "mindless material process." Behe, too, talked of "key scientific discoveries" that led one to conclude the truth of intelligent design. His indictment of evolution hinged on nineteenth-century drawings of embryonic fish, salamanders, chickens, and humans. Behe's final appeal was not to scientific authority but to a political one—the fact that his book, *Darwin's Black Box*, was reviewed in *National Review* by a biochemist. Fellow panelist Buckley had been the magazine's editor from its founding in 1955 until 1990 and was the founder of the *Firing Line*.[27] An appearance in 2007 on the PBS program *NOVA* provided another opportunity for Johnson. It was a science program, and Johnson used the language of science with talk of hypotheses, evidence, and experimentation. It was a choice between "two hypotheses," he said, "on the basis of evidence and logic."[28]

Johnson and Behe, had become the "other side" in evolution versus creationism, just as quoting a Democrat on a policy issue meant that a journalist also needed a Republican quote to balance the story. The two men made gains for their cause even if the coverage was negative because they were granted the status of intellectuals, academics, or scientists. Coverage kept alive the idea that a scientific alternative to evolution even existed, even when a story said there was no controversy among scientists. They elevated their cause when, in the public arena, intelligent design was associated with science, and Johnson and Behe were designated "intellectual challengers."

Scopes 2, 3, 4 . . .

One of the consequences of Johnson, Behe, and other academic creationist voices was to provide what creationists desperately needed in the wake of legal decisions against teaching creationism and intelligent design in public schools—science, or something that sounded scientific and did not refer explicitly to God or the Bible. *Edwards v. Aguillard* (1987) had necessitated the search for such a theory because the U.S. Supreme Court said creation-science was religion. The ruling had been a decisive loss for creationists. But it was not the end of legal challenges to teaching evolution in public schools, as some thought it would be. Even Stephen Gould wrote that the decision meant "creationists can no longer hope to realize their aims by official legislation."[29] The U.S. Supreme Court found Louisiana's equal-time law unconstitutional under the Establishment Clause and ruled that teaching creation-science was teaching religion. It was the end of equal-time laws, not antievolutionism. The decision necessitated some creative thinking and interpretation of the decision on the part of creationists, who had to find a new legal approach.[30] The Court did not rule on whether or not creationism was science. But the majority opinion said teaching different scientific theories about human origins could be done in the context of secular intent of advancing science education.[31] This was the opening for creationists, whose primary alternative became intelligent design. Creationists also cited the decision as an avenue for allowing the teaching of evidence against evolution.

Antievolutionists used a number of different approaches in state and federal courts, including arguments for freedom of expression or religion, disclaimers pasted in textbooks or read in class, and stickers that encouraged "critical thinking." Two aspects of court rulings in years following *Edwards* proved problematic for creationists: Teachers do not have a First Amendment right to teach creationism in public schools, and a district could prohibit teaching creationism, thus overriding individual free-speech rights based on a state's compelling interest in student education.[32] Such restrictions did not stem the flow of legal challenges.

Peloza v. Capistrano Unified School District in California (1994) arose when biology teacher John E. Peloza sued the district after he was fired for teaching his religious beliefs. In his lawsuit, Peloza argued that evolution was a religion.

Therefore, requiring him to teach it meant he was required to teach religion, a violation of the Establishment Clause. He further argued that evolution was not a valid scientific theory, and was one of two world views on the origins of life and the universe. The other view was creationism. The school's actions, he argued, prevented him from teaching his students the difference between "a philosophical religious belief system on the one hand and true scientific theory on the other." The 9th Circuit Court of Appeals affirmed the district court's ruling that evolution was not religion, and that the school's interest in avoiding an Establishment-Clause violation trumped Peloza's free-speech rights.[33] The decision did not deny that Peloza's free-speech right was compromised, only that it was superseded by Establishment-Clause concerns. The U.S. Supreme Court in 1995 rejected Peloza's petition.[34]

Arguing in a similar vein two years later, a substitute teacher in South Bend, Indiana, sued the school district after he was removed from the district's substitute list for, among other things, injecting religion into the classroom, reading the Bible aloud to high-school and middle-school students, and professing his belief in creationism in a fifth-grade class. The teacher, Peter Hellend, sued for discrimination based on religious beliefs. The U.S. District Court said Hellend's actions were a violation of the Constitution, and his firing was not discrimination based on religious beliefs. The court ruled that the district must disallow teachers from expressing religious views in the classroom, including creationism. In an interesting note, the court observed that in a letter to the school superintendent, "Hellend admitted that after he discussed creationism in the fifth grade science class, he told the students that he could get into big trouble for talking about Jesus and the Bible in school, and so he agreed not to assign homework if they would not tell anyone about the discussion."[35] The decision did not provide the source who revealed, presumably at risk of getting homework, Hellend's tactical pedagogy. Peloza and Hellend were defeats for creationists. However, creationists saw an opening in the cases, and that was to get religion or creationism per se out of the picture.

In *Freiler v. Tangipahoa Parish Board of Education*, Louisiana, the U.S. Fifth Circuit Court of Appeals affirmed a district court ruling in 1999 that it was unlawful to require teachers to read aloud a disclaimer before teaching evolution. The court said the disclaimer was intended to protect a particular religious view. Part of the disclaimer said teaching evolution was "not intended to influence or dissuade the biblical version of Creation or any other concept." The school district appealed to individual rights and argued that the disclaimer advanced "freedom of thought, as well as sensitivity to, and tolerance for, diverse beliefs in a pluralistic society." The court disagreed. The defense was a broadly cast appeal to deep cultural values: individual rights, freedom of religion and expression, and democracy. The court found, instead, that "the effect of the disclaimer is to protect and maintain a particular religious viewpoint, namely belief in the biblical version of creation." The U.S. Supreme Court later refused to grant certiorari to hear the case.[36]

The freedom-of-expression defense also fared poorly for creationists in 2000, in *LeVake v. Independent School District no. 656*. The Minnesota Court of Appeals ruled that a teacher's right to freedom of expression did not mean the teacher may teach material that circumvents the curriculum, and that not allowing a teacher to teach the so-called evidence against evolution was not a violation of free-speech rights. The teacher, Rodney LeVake, had spent one day on evolution in his tenth-grade biology class in the spring 1998 term. LeVake told the curriculum director he did not regard evolution as a "viable scientific concept," and in a position paper written for school administration, LeVake said he would teach evolution along with "difficulties and inconsistencies of the theory." He was reassigned to a ninth-grade natural science class for the next year. LeVake claimed to be a victim of discrimination because of his religious beliefs. The court found against him and his argument of individual freedom on the same basis of earlier cases: a compelling state interest in a suitable curriculum versus individual teachers' desires to "teach what they please."[37]

The Georgia Court of Appeals gave a slightly different twist to government neutrality on issues of religion. In *Moeller v. Schrenko* (2001), the court said using a textbook that denies the scientific validity of creationism is not a violation of the Establishment Clause. The case began with passages in a biology textbook that said the creation of life by a divine force could not be tested scientifically. A student sued, charging that such content denigrated her religious beliefs and free exercise of religion. The lower court found otherwise, ruling that the textbook and offending passages were neutral on creationism and did not advocate religious beliefs. The first of two offending passages said, in part: "A belief in divine creation, however, is not a scientific hypothesis that can be tested. . . . Whatever you propose, it is always possible to argue that a divine agent simply made things appear the way they do. . . . This is not to say that the belief is wrong, but rather science can never test it."[38]

The second passage at issue included the statement that science currently is not able to answer questions about life's origins.[39] The plaintiff argued the passages were not neutral because creationism was not elevated to science. Here was the Scopes-era objection come to new life: that science could be defined by individuals or majority vote but would not be constrained by empirical observation or other standards of the mainstream scientific community.

All of the above tactics—disclaimers, stickers, individual rights, redefining science—converged in another Georgia case in 2005. The U.S. District Court for Northern Georgia, in *Selman et al. v. Cobb County School District*, ruled it was unconstitutional to paste stickers in texts stating that evolution is a "theory, not a fact." The Cobb County Board of Education voted to put stickers in some science texts saying: "Evolution is a theory, not a fact, regarding the origin of living things. This material should be approached with an open mind, studied carefully, and critically considered." The parent-plaintiffs argued that the sticker promoted creationism and discriminated on the basis of religion and nonreligion. The court agreed, and said an "informed, reasonable observer would interpret the sticker to convey a message of endorsement of religion because it sent a message to those who opposed evolution for religious reasons that they were *favored members of the political community* [emphasis added]." In further recognition of the political context, thus political speech, at issue, the court said the sticker conveyed "an impermissible message of endorsement and tells some citizens that they are political outsiders while telling others they are political insiders." The court said the argument for promoting critical thinking was "not a sham," but other language in the sticker "somewhat undermines the goal of critical thinking by predetermining that students should think of evolution as a theory when many in the scientific community would argue that evolution is factual in some respects." The court said having no explicit references to religion weighed in favor of the sticker's constitutionality, but the board's primary purpose was not to promote critical thinking, but to accommodate religious beliefs "deemed inconsistent with the scientific theory of evolution." Again, the court said, the "informed, reasonable observer" would understand that the board specifically endorsed Christian fundamentalist and creationist views of evolution.[40]

That court was attuned not just to the political nature of the argument, but to the popular arena—versus professional, academic, or scholarly venues—in which the campaign was conducted. Here, the court said the use of the word *theory* was in a popular sense, and in the sticker the word meant "highly questionable 'theory or hunch.'" The court said this could cause confusion among students between theory and fact and detract from education, leaving less time to teach evolution and thereby diluting instruction to the benefit of antievolutionists.[41] The same use of language—"theory" as a guess, a hunch, speculation—had been causing confusion since before the Scopes trial, where Bryan had used the word to mean just that—a "guess."

In all of those cases, creationists lost the legal fight. However, the varied attempts to get around the *Edwards* ruling demonstrated how one could lose in court but win in public. The creationist arguments that failed in court—including freedom of speech, encouraging critical thinking, and various textbook disclaimers—were political winners because they resonated with people in a number of ways. First, the arguments made it sound as though there was an issue, and by implication two sides to the issue. Second, the appeals were to deeper cultural values, such as individual freedom and the right to rebel against the establishment in the name of that freedom. Moderation and accommodation are hallmarks of American political history,[42] and in cases of the 1980s and 1990s, creationists began recasting themselves in precisely that fashion. They were merely asking for equal time or offering an "alternative" view. Such a veneer of tolerance had the practical effect of making the opposition—mainstream science—look like the radicals or extremists because they would not move toward an imaginary middle and embrace religion as part of science.

Creationism's Nemesis

When Johnson read Dawkins's *The Blind Watchmaker* in 1987, Johnson was moved to write *Darwin on Trial* (1991). Inspired by one of Darwin's foremost defenders in contemporary times, Johnson lumped Darwin with Marx and Freud. All three, he charged, were guilty of employing reason without God to explain the world. Johnson called *The Blind Watchmaker* a "sustained argument for atheism," and noted Dawkins's remark that antievolutionists were "ignorant, stupid or insane." Johnson wrote, "Dawkins went on to explain, by the way, that what he dislikes particularly about creationists is that they are intolerant."[43] The remark exemplified an ability to win an argument politically, not scientifically, and not on the merits of the argument but by marginalizing the opposition. With Johnson's approach, Dawkins was the one who excluded people, while creationists were tolerant and inclusive.[44] Dawkins became the foil for conservative Christianity in general, creationists in particular, as he polarized and personalized the debate.

Dawkins has been a highly public advocate of science over the last several decades with numerous books, articles, speaking engagements, and media appearances. Dawkins made his case in his first three books: *The Selfish Gene* (1976), *The Extended Phenotype* (1982), and *The Blind Watchmaker* (1986).

The last one may have been especially egregious for creationists because Dawkins refuted William Paley's argument from design. Though published in 1802, Paley's *Natural Theology* is the basis of contemporary intelligent design, and so Dawkins was doing more than criticizing religion. He went after the intellectual core of creationism. Dawkins became the foremost pro-Darwin provocateur in the 1990s with *River out of Eden* (1995), *Climbing Mount Improbable* (1996), and *Unweaving the Rainbow* (1998). By the new millennium, he was the preeminent public voice not just for Darwinism but for atheism, which included harsh denunciations of religion, especially creationism. Two additional books, *A Devil's Chaplain* (2003) and *The God Delusion* (2006), expanded his essential premise. Plus, he appeared to revel in aggravating creationists. His battle with religion took on its sharpest edges in his remarks about antievolutionists. Dawkins demanded testable, falsifiable proof of God, which biblical literalists could not do because their basic premises are outside science, faith-based, and not amenable to empirical testing. This was the essence of creationists' fight with science: If creationists used scientific method to support their claim of God, they would falsify their own claim. Worse yet, creationists could not just move God outside science because doing so would be a denial of their basic premise. If book sales and media presence were any indicator, Dawkins was an effective communicator for evolution and science. He wrote clearly and convincingly, but he was trenchant in his condemnation of religion. His arguments for evolution may have appealed to the "middle" group, but any gains risked being offset by his strong antireligion feelings.

Dawkins's appearances on PBS tended to be slightly more politic than some of his writing, but he retained his edge and uncompromisingly condemned creationists. In *Interview on PBS with Bill Moyers*, Dawkins acknowledged that the issue was a very political one in America. "[N]ot a single member of Congress or the Senate would dare say that they don't believe in a supernatural God." A politician, he said, had to "pretend to believe in a supernatural God" whether or not he or she actually did. Dawkins was more compromising in that he said he was all for tolerance, and he averred that he, too, was religious in the sense that "God" was a literary personification of appreciating the "deep problems of the universe, and the things that we don't understand."[45] In another interview, Dawkins said Behe should "stop being lazy and should get up and think for himself about how the flagellum evolved instead of this cowardly, lazy copping out by simply saying ' . . . it must have been designed.'" Though Dawkins made his case scientifically, he left no room for anyone who may have been seeking a compromise, a political middle ground. "Darwin destroyed the argument from design," he declared.[46] His appeal would, in terms of practical politics, be much narrower than the creationist profession of inclusiveness, however vague.

Time called Dawkins the "foremost polemicist" for the idea that science was "chipping away at faith's underlying verities." The magazine arranged a debate between Dawkins and Francis Collins, director of the National Human Genome Research Institute, "forthright Christian," and author of *The Language of God: A Scientist Presents Evidence for Belief.* In a prelude to the debate, *Time* offered the issue in political terms: "[A]n argument in which one party stands immovable on Scripture and the other immobile on the periodic table doesn't get anyone very far. Most Americans occupy the middle ground: we want it all." Collins and Dawkins agreed that Genesis is not a book of science. They diverged from that point. Dawkins characterized fundamentalists as "clowns" and belief in God as a "cop out." Collins warned Dawkins that name calling would not help his case. Collins, by no means a representative of young-Earth creationists, said he did not find science and religion in conflict. Dawkins was implacable on the point: "What I can't understand is why you invoke improbability and yet you will not admit that you're shooting yourself in the foot by postulating something just as improbable, magicking into existence the word God."[47]

Dawkins spoke to a much narrower audience than creationists, who aimed for the general public and politicians, excluding no one, even though mainstream scientists were unlikely to be converted and courts have yet to be swayed. But the latter two groups were never excluded. Dawkins, by contrast, cast out the believers and left them no room to even peer into the windows of scientific inquiry. This did not mean Dawkins was wrong in his assertion that God was not scientifically knowable. It did mean his approach was less effective politics. Creationists had a more adaptable marketing strategy that embraced a larger number of people.[48]

Compared to the creationists' moderate tone, Dawkins sounded radical, even menacing. It was a role reversal from the science and religion personas of *Inherit the Wind.* The creationist appeal to tolerance, critical thinking, equal time, fairness, and so forth was set against Dawkins, who wrote repeatedly that God does not exist, religion is myth, and science eventually can explain everything. His position is philosophically daunting, especially for individuals who are not dedicating their spare time to pondering the mysteries of life's origins and diversity.

By comparison, creationists, such as Johnson, did not demand much intellectually. You may accept science, they said, just as we do. You may even accept certain aspects of Darwinian evolution. There are issues, they admitted. But the only real dictate was to accept the premise "God exists." Then, you are one of us, whether you knew it or not. This approach often will win the middle ground and appears much more tolerant than Dawkins. The creationist approach ceded some ground to old-Earth creationists, even theistic evolutionists, but maintained and strengthened the political foundation. In this respect, creationists had learned from Darrow and from *Inherit the Wind*. Radicals and extremists are unappealing. Rebels are fine, even admirable, as long as they adhere to certain core cultural values. Dawkins was in the intellectual, contrarian tradition of Darrow, but he was not inclined to appeal to the masses or compromise his position for the sake of his cause. No American politician would have quoted him. Undeniably, part of the appeal of Dawkins to the news media was his radicalism. Being so far from mainstream religion made him unusual and, therefore, even more newsworthy.

Creationists acknowledged, even welcomed, Dawkins. In *Darwin's Nemesis*, Dawkins was the antithesis of the volume's subject, Phillip Johnson. Contributors used Dawkins to validate their complaint of exclusion. Evolution, Dembski argued in the volume, could not stand up to scrutiny, and evolutionists endangered their arguments by the very act of engaging antievolutionists. "Richard Dawkins is a case in point. Dawkins refuses, as a matter of principle, to debate me and my colleagues because it would, in his view, dignify our position. Yet he cannot resist criticizing us in print. Notwithstanding, whenever he does so, he makes himself vulnerable."[49] The point may be an inadvertent acknowledgment of the fact that in politics it is much better to be excoriated and denounced than to be ignored. The latter suggests insignificance.

The political advantage to the creationist approach could be seen in the response to Stephen Gould's 1999 book, *Rocks of Ages*, in which he introduced the concept of "non-overlapping magisteria" (NOMA). Science and religion, he wrote, were different ways of knowing, but were mutually exclusive and should be approached as such. Gould, who called himself an agnostic, offered the idea in the spirit of compromise between the two realms, without ceding absolute authority to either one.[50] Both sides lambasted him. Creationists criticized Gould's separation of religion from nature, a sort of theological apartheid. Dawkins accused Gould of "bending over backwards to positively supine lengths." NOMA, Dawkins said, "sounds terrific—right up until you give it a moment's thought." Dawkins believed NOMA excused religion from intellectual scrutiny.[51] In an interview published in *Time* magazine, Dawkins said, "I think that Gould's separate compartments were a purely political ploy to win middle-of-the-road religious people to the science camp. But it's a very empty idea. . . . Any belief in miracles is flat contradictory not just to the facts of science but to the spirit of science."[52]

Susan Jacoby, historian of freethinkers in America, wrote that Gould's "accommodationism" had a long tradition in Protestant America, and such an approach "had the effect of making agnostics, atheists, and uncompromising rationalists look like crackpots and extremists—an image that endures to this day."[53] This would work to the advantage of creationists as they moderated their tone while they offered up Dawkins and Dennett as representatives of anyone who opposed teaching an alternative, which meant intelligent design or creationism, in public schools.

Mere Creation

In 1996, intelligent-design advocates gathered at Biola University, a Christian university in Southern California of about 6,000 students, for a conference dedicated to promoting their research. Its significance was not in the research, because no peer-reviewed publications resulted, but in the creation of an ID-campaign strategy, which came to be called the "wedge strategy."[54] The Discovery Institute's Center for the Renewal of Science and Culture was the conference's chief sponsor and prime mover behind *The Wedge Document*, which grew out of the conference.[55] *The Wedge Document* was sophisticated and well considered, clear as to goals, timelines, and constituents. The issue was the same as in 1925, the threat of modernism to creationism. Evolution represented the former, intelligent design the latter.

In *Mere Creation: Science, Faith & Intelligent Design*, a collection of papers from the conference, Johnson cited three seminal events in the second half of the twentieth century that symbolized what he believed was an ideological shift. The first was the 1959 Darwin centennial, which commemorated publication of *On the Origin of Species*; the second was the 1960 movie *Inherit the Wind*; the third was the 1962 Supreme Court decision, *Engel v. Vitale*, that banned prayer in schools. Johnson said these events changed the nation's "ruling philosophy," replacing biblical morality with a "purposeless material process." According to Johnson, *Inherit the Wind* created the stereotype that closed minds to the place of the Bible in scientific inquiry, an assertion he had made five years earlier in *Darwin on Trial*. *Mere Creation*'s postscript by Bruce Chapman, president of the Discovery Institute, said the Scopes trial was important because it "foreshadowed decades of legal theory that rationalized outrageous and irresponsible behavior. . . . This is not to say that science per se was responsible for the demoralization of our culture.

Rather it was the materialist interpretation of science."[56] Dembski, Chapman, and Johnson were arguing the Scopes case—more than seventy years after the fact. Though deemed a science conference, it was organized like a political convention: identify and affirm the issues; state and affirm the policy; chart a course of action to achieve policy goals or to win election. The last point was the real accomplishment because *The Wedge Document* was a campaign plan.

Contributors to *Mere Creation* touted the conference as a breakthrough event, one that brought together creation-scientists to provide what they saw as a scientifically legitimate alternative to materialistic science. Communication was stated as a major goal of the conference. In fact, communication seemed inordinately large for a group whose professed primary goal was to win scientific legitimacy. Again, the path to legitimacy was not via science, but publicity. *Mere Creation* anticipated such doubts and denounced the exclusivity of mainstream science and the exclusion of any alternative that involved a creator. But it was not just any creator. It was the Christian God to which they dedicated themselves, made clear in the opening paragraph of the Introduction, by Dembski. The clarion call was a political one, a big-tent approach couched in a way to invite a broader view, not just young-Earth creationists, but anyone who believed in God: "There is, however, an alternative approach to unifying the Christian world about creation. Rather than look for common ground on which all Christians can agree, propose a theory of creation that puts Christians in the strongest possible position to defeat the enemy of creation, to wit, naturalism."[57] The alternative was intelligent design, which Dembski deemed compatible with "everything from utterly discontinuous creation (e.g., God intervening at every conceivable point to create new species) to the most far-ranging evolution (e.g., God seamlessly melding all organisms together into one great tree of life)."[58]

Mere Creation's eighteen chapters were important to the political agenda because the culture to which they appealed was, and is, enamored of science and technology but often is leery of or hostile to the philosophic implication of a science that is indifferent to the God hypothesis in that it is not testable. Dembski's Introduction and Johnson's "Afterword: How to Sink a Battleship: A Call to Separate Materialist Philosophy from Empirical Science" explained to readers how they can have both God and science, contrary, they stated, to what materialistic scientists such as Dawkins proclaimed. *Mere Creation* and intelligent design were the scientific patina that would appeal to a culture that loved science but was indifferent as to how science worked. Here was an idea that conveyed scientific respectability while embracing religion and accommodating ignorance.

Those writing in *Mere Creation* used Dawkins as a foil. Dembski declared to readers in the Introduction that Dawkins and Dennett were guilty of "sheer arrogance" because they charged ID theorists with being "stupid or wicked or insane for denying the all-sufficiency of undirected natural processes in biology" and compared "challenging Darwinism with arguing for a flat earth." Ironically, the first words in the book's first chapter were "Richard Dawkins," who was assailed for his 1996 book *Climbing Mount Improbable*. Here was, for ID proponents, their antithesis—defining themselves by what they were not.[59] So there was a certain literary neatness to conclude the volume with Dawkins. As in the beginning—of *Mere Creation*, that is—the volume skewered Dawkins for arrogance, "bluster," and materialism.[60] And lest readers forget themselves, the Afterword, by Johnson, demonstrated the virtue of humility: "Some of us saw a clip of Richard Dawkins being interviewed on public television about his reaction to Michael Behe's book. You can see how insecure that man is behind his bluster and how much he has to rely on not having Mike Behe on the program with him, *or even a lesser figure like Phil Johnson* [emphasis added]."[61]

They had found in Dawkins an opponent who, like Darrow in the defining event for creationists, was eloquent, logical, and dedicated. But Dawkins lacked the folksy, suspender-snapping charm of American history's favorite village atheist.

The Wedge Strategy

The Wedge Document formalized creationists' campaign goals and tactics, provided a working document for coordination of organizations that were far-flung, not just geographically but also in theology and politics. It was an artfully constructed campaign platform and strategy outline and a model of clarity and logic—in the context of the goals of the Center for the Renewal of Science and Culture. The document said the proposition that God created man was at the foundation of "democracy, human rights, free enterprise, and progress in arts and sciences." With that established, those who were at the heart of tearing down this idea were listed: Karl Marx, Sigmund Freud, and Charles Darwin. They were castigated for materialistic thinking that denied spirituality, morality, and personal responsibility. *The Wedge Document* said it sought "nothing less than the overthrow of materialism and its cultural legacies," and a reopening of "the case for a broadly theistic understanding of nature." A three-part strategy included projects, a five-year strategic plan, and goals for the next five and twenty years. The three-phase plan provided objectives and timelines and stated that many of the goals in the first two phases could be accomplished within the first five years (1999–2003). Phase one was research, writing, and publication. At this point, according to the plan, it was not necessary to have superior numbers because, "Scientific revolutions are usually staged by an initially small and relatively young group of scientists who are not blinded by prevailing prejudices." This stage was described as persuasion—not indoctrination—via "solid scholarship, research and argument." With that foundation, phase two aimed at promoting popular reception of theistic science. The document stated that in phase two the tactic was to cultivate print and broadcast media contacts, think-tank leaders, scientists, academics, congressional staff, talk show hosts, and other opinion leaders. The connections to media and politics, according to the plan, prevented the wedge from being "merely academic."

The authors did not lose sight of their goal: popular support, not scientists or academics. "We intend . . . to encourage and equip believers with new scientific evidence that support[s] the faith, as well as to popularize our ideas in the broader culture." According to *The Wedge Document*, phase three logically followed the first two phases of creating a body of writing and scholarship, and priming the public for accepting intelligent design. Now, it was time for "confrontation with the advocates of materialist science." This meant getting legal help in making intelligent design part of public-school curricula. According to the plan, this should "draw scientific materialists into open debate with design theorists, and we will be ready."[62]

The first of three twenty-year goals was to make ID the "dominant perspective in science." It may have sounded outlandish, but it was not, if one considers young-Earth creationism's wide acceptance in American culture. The idea has persisted since the 1920s at the state and local level, and was endorsed by a recent U.S. president and numerous candidates for that office. The second goal was to entrench intelligent design in specific fields, including biology, physics, cosmology, psychology, theology, and philosophy. Not losing sight of the means to such an end, goal three was "to see design theory permeate our religious cultural, moral, and *political* life [emphasis added]." The movement gave traction to its mission with very concrete, measurable goals over a five-year period—thirty books on design and its cultural implication; one hundred scientific, academic, and technical articles; "significant coverage in national media," such as *Time*, *Newsweek* and PBS; ten states with design in public-school curricula. The way to achieve such goals, the document stated, was to build alliances in academe, media, religious groups, social-advocacy groups, and so on. *The Wedge Document* cited scholarly research, but committed itself also to opinion columns, documentaries, and other "popular" avenues.[63]

In all, it was a pragmatic plan with concrete, achievable goals—if the goal was political gain rather than scientific knowledge. *The Wedge Document* recognized the importance of scientific legitimacy, but stressed campaigns in popular media rather than peer-reviewed research in the scientific community. The contrast between the literalists of the 1920s and the creationists of 2000 were sharp in some respects, and indistinct in others. The core issue was opposition to modernism and evolution, and the insistence on the Bible as the preeminent authority for all knowledge. The decades were distinctive for creationists because the movement had become more organized and more widely accepted, and even gained recognition as the "other side" in a national debate. In the 1920s, antievolutionists were united tenuously around the doctrine of literalism and separated by denominationalism. By the 1990s, creationism had been rebranded and liberalized, embracing far more than young-Earth creationists, and it proponents were organized around sophisticated campaigns, institutes, and publications.

5

Science on Trial
The Ghost of Bryan

We have the purpose of preventing bigots
and ignoramuses from controlling the education
of the United States.
—Clarence Darrow, July 20, 1925

I am simply trying to protect the word of God against the
greatest atheist or agnostic in the United States.
—William Jennings Bryan, July 20, 1925

Bryan's twenty-first-century resurrection occurred in two places in the heartland. The first was Kansas, just south of his native Nebraska, the second a small town in Pennsylvania. In both cases, Bryanesque rhetoric permeated fervent appeals to individual rights and democratic principles.[1] One side fumed for science and against theocracy. The other side railed about the assault on religion and bemoaned the abandonment of sacred traditions.

"There are two worldviews that are in conflict." The line could have come straight out of the Scopes trial, from either Bryan or Darrow. But it was October 2005, and it was the *New York Times* quoting an attorney for the pro-creationist school board members in Dover, Pennsylvania. Another attorney, in a similar California case, put the contest more starkly: "The contest between evolution and Christianity is a duel to the death, between the unbelief that attempts to speak through so-called science, and the defenders of Christian faith."[2]

The intelligent-design advocates were in the front lines and the headlines of the national "culture war," which had become a common term in commentators' lexicons and applied to everything from presidential elections to birth control. Darwinism was in the middle of it all, according to the *Washington*

Post: "In an escalation of the nation's culture war over the teaching of evolution, the National Academy of Sciences and the National Science Teachers Association announced yesterday that they will not allow Kansas to use key science education materials developed by those organizations."[3] In another story, the *Post* reported that a group was taking on the religious right in a campaign that began with "trying to learn whether 'regular citizens could take effective action to counter the Cultural War initiated by the leaders of the Religious Right.'"[4] In Harrisburg, Pennsylvania, the *Patriot-News* stated, "The (Dover) trial is a culture-wars phenomenon" and a columnist said the schools were "pawns in the culture war." Intelligent design was politics red in tooth and claw and had become a line of demarcation in the cultural conflict.[5]

Dover and Kansas

In Topeka, Kansas, the State Board of Education adopted science standards in 2005 that treated evolution as a flawed theory. The new standards redefined science to encompass supernatural explanations, presented intelligent design as an alternative to evolution, injected into curricula allegations of scientific controversies about evolution, and required that students be told evolution is a theory and not a fact. Public hearings on adopting new antievolution standards began in May 2005. Those hearings, staged like candidate debates, provoked press coverage and the indignity of scientists, who boycotted the hearings. More significantly for creationists, the press showed up. Scientists had been politically outmaneuvered because if they took part in the hearings, they gave credibility to creationists' assertion that a scientific controversy even existed. If they shunned the event, scientists were susceptible to accusations of arrogance or elitism, and of not having a good answer to critics. Creationists found an attorney to stand in for the evolution side. This made it possible to have a media event—two sides, disagreements, drama, all the necessary ingredients for getting on the news. But the attorney, Pedro Irigonegaray of Topeka, agreed only to a limited role. He would not testify for evolution, but would represent mainstream science by questioning the creationist expert witnesses. The hearing was reminiscent of the Scopes trial in that its outcome was a foregone conclusion, and the goal was to win publicity and create a press carnival. As in the Scopes trial, many national media left the Kansas event early. Besides, the press had the story. It was akin to the press exodus from Dayton before the famous Bryan-Darrow confrontation in the courtroom. In Kansas, the state board had its "evidence" thanks to a stream of creationist/antievolution testimony. One board member bemoaned the lack of evolutionists' testimony. In her lament about the creationist side being charged with close-mindedness, she asked "which side of the issue is being close minded? Why are some scientists tenaciously holding onto the evolutionary tenets that are unproven, as we have heard, and are often disproven?" The director of the National Center for Science Education said it had been a "show trial, like in the sense of the Soviet Union back in the fifties. The board already had its conclusion. They're just going through these motions, making a big show for the public, to get an idea out."[6]

In November 2005, the board gave final approval to the standards, which supporters said would permit an open, healthy debate about the strengths and weaknesses of evolution. The Discovery Institute and the Kansas Intelligent Design Network (KIDN) had been strong supporters of the standards and the campaign to adopt them. The Discovery Institute, founded in 1990 in Seattle, Washington, is a conservative Christian think tank dedicated to promoting creationism and fighting materialism—evolution, specifically. Though it has a number of "fellows" who work on intelligent-design research, its primary function—and accomplishment—is political. It is dedicated to winning public acceptance of its alternative to evolution. Intelligent design or young-Earth creationism, that is. The KIDN joined the Discovery Institute to propose more than twenty pages of revised science standards for Kansas. The state's Board of Education had appointed a science standards committee, composed of educators and scientists, to review the standards. The committee rejected all of KIDN's proposed revisions.

The Kansas crusade was a showcase event for the effectiveness of the Discovery Institute's wedge strategy. The long-term goal of defeating evolution and replacing it with biblical literalism was, with the "wedge" approach, subjected to the tactical reality of winning politically what creationists were losing scientifically. The goal was to defeat "scientific materialism" and replace it with "the theistic understanding that nature and humans are created by God."[7] To that end, teaching the controversy was an important component in the wedge strategy's first stage, which was criticizing evolutionary theory. This early phase of the campaign simply created or enhanced doubt about evolution in the guise of being fair or democratic in the approach to information. So Kansas did not outlaw evolution, but just demoted it to one of several options. It did not require Christian schooling, but only the chance to study "alternatives" to materialistic science. In this respect, it was very good public relations because the creationists were not objecting to anything, only seeking the inclusion of more possibilities, sounding as if critical inquiry was at the top of their agenda. In 2006, Kansas voters threw out four of the six board members who supported the intelligent-design–friendly standards. The new board rescinded the standards in February 2007 and again defined science as "natural explanations."[8] In Kansas, for the moment, the fight was resolved at the ballot box. In Pennsylvania, it was settled in federal court.

The Dover conflict began in 2004 with school board members who were young-Earth creationists wanting to promote their belief that life's complexity was scientific proof of an intelligent designer. On the other side, a number of parents in the school system saw teaching intelligent design as promoting religion in public schools. The legal issue began to take form in August 2004 when the school board accepted a donation of several hundred copies of the book *Of Pandas and People*, which would not be used directly in science class but would be made available as a reference for students. *Pandas* advocated intelligent design and criticized evolution. In October 2004, the board voted to require ninth-grade biology teachers to mention intelligent design as an alternative to evolution and to inform interested students about the availability of *Pandas*. Eleven parents announced a lawsuit on December 14, 2004, in Harrisburg, Pennsylvania, based on the First Amendment's Establishment Clause. In January 2005, the board directed teachers of ninth-grade biology to read a statement to the class before teaching the section on evolution. The statement read, in part:

> Because Darwin's theory is a theory, it continues to be tested as new evidence is discovered. The Theory is not a fact. Gaps in the Theory exist for which there is no evidence. . . .
>
> Intelligent Design is an explanation of the origin of life that differs from Darwin's view. The reference book, *Of Pandas and People*, is available for students who might be interested in gaining an understanding of what Intelligent Design actually involves.
>
> With respect to any theory, students are encouraged to keep an open mind.[9]

It was rhetorically masterful, even if the science behind it was not. In a short statement, it managed to accommodate and diminish evolution without denouncing it. It called for keeping "an open mind," which was an implied criticism of what the creationists saw as close-minded mainstream scientists. The statement promoted individual initiative in the guise of encouraging students to simply check out the alternative for themselves. Some parents saw it differently. Those eleven parents, including Tammy Kitzmiller, the case's namesake, thought the statement and *Of Pandas and People* injected religion into the curriculum. Kitzmiller was the mother of a ninth-grader and believed it was her province, not the school's, to discuss religion with her children. When the lawsuit was announced in Harrisburg, she was among the plaintiffs, attorneys, antievolutionists, and journalists on hand. She and other parents addressed the gathering. Kitzmiller said the policy was wrong. "The Dover school board created this policy for religious reasons."[10]

The ID advocates' political machinery outran their intellectual machinery. The Discovery Institute was drawn reluctantly into the case by its own *Wedge Document* adherents, who were pushing to engage the "enemy." But the DI was not ready. On the same day the suit was filed against the Dover school board, the DI issued a statement that called the local policy "misguided" but commended its attempt at "trying to teach Darwinian theory in a more open-minded manner," i.e., including the intelligent-design alternative. It asked the board to withdraw and rewrite its policy, but to no avail. The DI could not control the language and terms of the Dover debate and, thus, saw a losing case looming. It was a conundrum for DI because not participating meant refusing to support its own agenda at the local level, and participating appeared to mean taking the losing side.[11]

The case went to federal court in September 2005. The DI's concerns were borne out in the trial, and illustrated especially well at two points in the proceedings. The first was the plaintiffs' dissection of *Pandas*, which was at the center of Dover's policy. The second was the testimony of Michael Behe. On the first point, the scalpel testimony came from Barbara Forrest, coauthor of *Creationism's Trojan Horse: The Wedge of Intelligent Design*. She showed that in prepublication drafts of *Pandas*, the word *creation* and its cognates were used extensively. But after the Supreme Court's 1987 decision in *Edwards v. Aguillard*, the Louisiana case that went decisively against creationists, *intelligent design* replaced *creation* terminology in *Pandas*. Nothing else was changed. In addition, the definition of *intelligent design* in the 1987 edition of *Pandas* was the same as the definition of *creation* in the prepublication drafts.[12] In spite of the absence of explicitly religious language in the book, *Pandas* was a creationist text.

As further evidence of intelligent design's religious foundation, Forrest cited a 1992 conference concerning the wedge strategy in which a fellow of the DI's Center for Science and Culture described the participants as "a new breed of young evangelical scholars," which included Behe and Phillip Johnson. The former came to the stand in Dover as a premier biologist for the intelligent-design cause. Behe defended intelligent design and his personal contribution to the idea, which was "irreducible complexity." Behe denied that intelligent design was religious. But his definition of science proved problematic for his cause: that which "relies exclusively on the observable, physical, empirical evidence of nature plus logical inferences." It was the last part that got him into logical trouble. Anything natural processes could not explain must, by "logical inference," be explainable only by a "designer." Upon cross examination, he admitted that under his definition, astrology would qualify as a science. In another telling moment, the plaintiffs' attorney read from *Darwin's Black Box* that no scientific answer existed that could explain the immune system. The attorney produced more than fifty articles and books on just that subject. Behe said he had read none of them.[13]

In November 2005, voters ousted eight of the board members at the center of the decision to teach intelligent design as science. In December, federal district Judge John E. Jones ruled that teaching intelligent design in public schools promoted a specific religion and violated the First Amendment's Establishment Clause. In a 139-page opinion, Jones lambasted the board for its "breathtaking inanity" as he placed intelligent design in its historical context of creationism and antievolutionism. Among other things, he cited Behe's inability to differentiate between creationism and intelligent design, which, the judge wrote, is creationism, and, therefore, is religious. "[T]he writings of leading ID proponents reveal that the designer . . . is the God of Christianity." Furthermore, "ID is nothing less than the progeny of creationism."[14]

The reaction to the decision revealed an expectation of a politically driven decision. The press saw the story as a political one as much as one about religion and the First Amendment. Repeatedly, press accounts reminded people that Jones's blistering critique of intelligent design was from the hand of a Bush appointee to the federal bench—implying surprise that a Republican could find against creationism. Or, at the least, there was some mild astonishment that a Republican appointee could find against a strong Republican constituency. The *Los Angeles Times* said it most succinctly: the George W. Bush–appointed judge rendering this decision was a "church-going conservative."[15] The *New York Times*' background to the decision said that Dover usually voted Republican, but in the November elections had removed the school board members who advocated intelligent design and elected Democrats in their place. The *Washington Post* placed the trial as the latest skirmish in the "centuries-long cultural war." The *Post* included the Dover conflict in an analysis of the political woes of incumbent Pennsylvania Sen. Rick Santorum, a Republican. He had withdrawn affiliation with a Christian rights law center that defended the Dover schools policy, but he had earlier praised Dover schools for attempting to teach the controversy. Santorum also was on the advisory board of the Thomas More Law Center, which aided the defense in Dover.[16] He appeared to be caught between common sense and some constituents. It ultimately was for Santorum a political problem, not a scientific one. (He decisively lost his reelection bid in November 2005, winning only 41 percent of the vote. He reappeared in the 2011–2012 Republican presidential primaries.)

A few days after Jones's decision, NPR aired an interview with Forrest. Going to the political heart of the issue, reporter Ira Flatow asked: "Were you surprised that this was a judge who considered himself to be a conservative Republican and appointed by President Bush, yet he did not agree with the president on this?" Forrest did not answer the question directly, but said she was happy with the decision. The interview showed the inability of the event to escape the cultural myth, as Flatow noted the trial's parallel to *Inherit the Wind*.[17] Because NPR is public, rather than private, it was obligated to work harder to find a center. NPR had aired a number of episodes critical of intelligent design. Even though the trial was both a science issue and a political one, the latter made finding a center possible. As a purely scientific issue, it would have been very difficult, if not impossible, to find such a center where one could mollify the majority.

Scopes's Shadow

The recasting was a bit muddled, but the storyline was true to form. The press and public were ready for Dover. For decades, "Scopes 2" had been cultural shorthand for science-religion conflicts in public schools. Behe assumed the mantle of Bryan, if a bit clumsily because he lacked the national recognition and the oratorical skills. But he ascended the witness stand to defend his faith, and was defeated indecorously by the plaintiff's attorney, Eric Rothschild. Like Bryan, Behe looked bad in the press accounts, probably changed no minds, and only affirmed opinions on each side. Rothschild played Darrow's role well, even with a point of ridicule when he forced Behe to admit that his definition of science would accommodate astrology. As in the Scopes trial, outsiders played a big role: Darrow and the prosecution invaded Dayton, Tennessee, in 1925; ACLU lawyers and witnesses from afar converged on Dover in 2005. In both cases, the locals ended up looking like buffoons, even duplicitous ones in Dover, where board member Bill Buckingham's testimony only added to ID's image agonies. He said in deposition in January 2005 that he did not know who funded the donated copies of *Pandas* to the schools. But his testimony at the trial showed he appealed to his church for donations for the books so tax dollars would not be used. He raised about $850. He wrote a check for $850 from his own account, made out to Donald Bonsell, father of a fellow board member, Alan Bonsell, also an ID advocate. Donald Bonsell, Alan later admitted under oath, donated to the board the money for purchase of *Pandas*. In the trial, Buckingham denied lying in his deposition. But Jones, in his decision a few weeks later, found that Buckingham and Bonsell had lied under oath in their January depositions about the source of funds for *Pandas*.[18]

In Dayton, local businessmen had contrived the Scopes trial for publicity, which turned on the town and its people in the long run. First, H. L. Mencken verbally lacerated the locals as hillbillies and rubes. Later, *Inherit the Wind* imprinted in cultural legend the image of the locals as narrow-minded mob-driven bigots, menacing in their ignorance. In Dover, the "insiders" were a few school board members who decided their truth was the truth. There was no Mencken to make them appear fools, but several decades of court decisions and press accounts, inevitably recalling Scopes and *Inherit the Wind*, had primed the national audience to see theocratic zealots on the side of religion. Creationists seemed fitted for the stereotype. In Scopes's case, the final disposition came from the state supreme court, which overturned the conviction in what it deemed a "bizarre" case. In Dover, Jones found the board's actions to be inane and deceptive. In the Scopes tradition, the drama was a fairly simple one: two sides, heroes and fools, modern and primitive, moral and immoral, and nothing less than the future of civilization at stake— the casting always depending on one's inclinations in the literalism versus evolution fight.

In Dover, the networks simplified it best for America. *ABC Nightline*, in an episode titled "Science vs. Religion," opened with a clip from the *Inherit the Wind*, and so affirmed the historical continuum. For *ABC World News Tonight*, the war metaphor dominated several episodes on the evolution issue, and *ABC News* could not help but present ID as an "alternative," with the trappings of a "theory." In its earliest episode of 2005 coverage of the Dover case, ABC set a Church of Christ pastor against an ACLU attorney. Reporter Dan Harris introduced the story as the "latest, fiercest front in the nation's culture wars." The episode featured the Museum of Earth History in Eureka Springs, Arkansas, where the institution's founder—Thomas Sharp—said on camera that it was "biologically impossible" for the human brain to have evolved from slime. *ABC News* reporter Jake Tapper pointed out that the goal of the museum was to "put dinosaurs and other scientific discoveries in the context of Noah and the great flood and Adam and Eve."[19]

CBS said Dover was a part of the political and cultural war—belief in God versus violation of church-state separation. In May 2005, CBS heightened the culture-war theme by pairing Kansas and Dover with a story on homosexual marriage. Reporter Bob McNamara introduced the story as a "war of words" concerning "creation theories," and gave the president credit for sharpening the conflict: "Evolutionists say, President [George W.] Bush's reelection gave Christian conservatives a new momentum on morality issues, and science teachers especially feel the heat."[20] It neatly combined science, politics, morality, and religion, as it elevated creationism to the level of theory. NBC approached the story as it would a liberal-conservative debate on national defense or social policy. Intelligent-design advocates often became "conservative Christians" or Republican challengers to status quo. Following the structure of an election story, NBC reported poll results showing 57 percent of Americans believe in the biblical account of creation.[21]

ABC's *Nightline* devoted three episodes to the intelligent-design story in 2005. The January 13 edition opened with several community members in Dover declaring intelligent design a religion, not science. They accused intelligent-design advocates of wanting "public schools to operate like Sunday schools." This was set against two other citizens, one stating, "Evolution itself is a religion. And it's an atheistic religion." *Nightline* host Ted Koppel cited a Gallup poll showing 35 percent of Americans believed the facts supported evolution, while 35 percent did not believe so, and 29 percent did not know. The show, "War in Dover," described a small town "torn up by an argument over an idea," and a school board "under siege."[22]

The Dover trial was a political story about the culture war, the "conservative Christians" or "conservative Republicans" versus anyone who opposed them. In the news stories about the trial, intelligent design often was one of a number of issues that defined conservatism, which also included favoring prayer in schools. The press would be nonpartisan, or "objective," by simply reporting each side, being fair and balanced. Such a two-sided approach made the trial a two-party political contest rather than a religious issue. Journalists lived up to professional values of fairness and balance, which fit easily into the broader political-story template. In defense of the press coverage, such an approach made the story comprehensible to the larger audience by simplifying it, making it meaningful to everyday lives, and providing the familiar historical and political context.

The versatility and usefulness of the political template was illustrated very well in a CNN *Crossfire* episode, "Can Evolution and Intelligent Design Be Taught Together?" The cohosts were James Carville and Robert Novak. Both are nationally known political commentators, Carville as lead strategist for Bill Clinton's presidential campaign and Novak as a political columnist and commentator. Notably, neither offered expertise in science or theology. During the May 2005 show, various political constructions were heaved into the science-religion debate. Novak said it was a matter of allowing "other theories" to be taught, and not doing so was "intolerant." He alluded to a CNN/USA Today/Gallup poll showing 76 percent of people would not be upset if creationism were taught in public schools. This resurrected Bryan's distortion of Jeffersonianism. Such a view meant majority rule validated ID, whether or not it was scientifically valid.[23] But it also showed the way in which the scientific community had lost control of the debate. Creationists still lost consistently in court, but won with much of the public.

The Campaign

That an issue even existed was testament to the success of the creationist wedge strategy. Merely winning time on national media and space in newspapers was validation even if it was only by virtue of publicizing the grievance, without regard to its intellectual merit or lack thereof. Giving Dover's intelligent designers a national platform suggested significance. Most people in the audience would not be inclined to sort through the foibles of news production in order to understand how creationists manipulated news values. But people would understand a good fight, which had a history, with winners and losers.

News coverage legitimized the issue in a political context. Creationists became conservatives, and opponents became moderates or liberals, sometimes implicitly, other times explicitly. An August 2005 *Nightline* episode, "Doubting Darwin: The Marketing of Intelligent Design," said it was an old debate, in which the issue may be more political than scientific. Reporter Chris Bury cited President Bush on several occasions and his endorsement of "teaching the controversy." But in spite of such a critical introduction, Bury deemed ID an alternative to evolution and the show balanced perspectives: a Case Western Reserve University professor who called teaching the controversy a "brilliant" public relations tactic, versus a Discovery Institute fellow talking of evidence, theory, and the designer. To accentuate the political nature of the debate, two conservative commentators were part of the fray. Columnist Cal Thomas accused scientists of arrogance by refusing to debate the issues. He grouped the issue with abortion, school prayer, homosexual marriage, and the exclusion of "God-fearing, tax-paying . . . patriotic Americans" from the wider culture. Columnist George Will acknowledged that marginalizing certain groups was a legitimate concern—but not a scientific issue.[24]

In the Kansas case, it was politically conservative or moderate designations, which corresponded to an individual's position on the proposed science standards. The *Kansas City Star* reported on an upcoming vote on science standards:

> As anticipated, the Kansas Board of Education approved its science standards with a 6–4 vote that fell along the board's conservative-moderate split. . . .
>
> Supporters . . . say the changes will allow for an honest discussion about the strengths and weaknesses of evolution, which they say is accepted with blind faith by mainstream scientists.[25]

A May 2005 *New York Times* story concerning the Dover case referred to the primary election as being "closely watched across the nation because of its implications for the contentious debate over evolution." "Christian conservatives" opposed teaching evolution in schools.[26]

A *Washington Post* editorial said the ruling in Dover "wounded a politically influential movement" and added that the decision would "not cast them into the political and cultural wilderness" because they had brought intelligent design to "the center of legislative debates in more than a dozen states." The *Post* noted that the president of the Southern Baptist Convention's Ethics and Reform Liberty Commission was a political ally of White House advisor Karl Rove.[27] A few days later, the *Post* emphasized Judge Jones's appointment by Bush, setting that in sharp contrast to quotes from Jones's ruling that ID was a mere relabeling of creationism and was not science.[28] The *Atlanta Journal-Constitution* also played it as an election story, in which the "voters issued their own verdict" in Dover, where "they went to the polls last week and threw out eight school board members who forced creationism into the curriculum."[29]

Locally, the *Harrisburg Patriot-News* said the conflict was not just a local school-board fight, but also was raging in the state legislature and in school boards across the nation. It was a "matter of fairness," according to some, as well as minority rights to an opinion about something that is "just . . . a theory." Others called it a "scientific inquisition" versus those who labeled it a "religious inquisition."[30] A feature story on Jones in the *Patriot-News* devoted substantial space to his political background and, like other media, saw Jones's appointment by Bush—a political act—as an important part of the intelligent-design story.[31]

The format of television news, with fewer words and more visuals than newspapers and magazines, necessitated some verbal shorthand. So intelligent design became "conservative groups."[32] NBC anchor Tom Brokaw introduced one extended feature on evangelical Christians by acknowl-edging their power in American culture and politics. The story was about various aspects of modern evangelism and the culture wars: From "battling sexuality in the media, to challenging the scientific notion of evolution, to fighting to display the 10 Commandments in government buildings, re-ligious conservatives are front and center in what they see as an ongoing war over the culture of America."[33]

CBS Evening News, in a brief story on December 20, 2005, about Jones's ruling, noted that it was a Bush appointee against intelligent design, and that the school board responsible for the policy had been voted out of office in November. It was political drama: "a small town in rural Pennsylvania. Mostly white, mostly Christian, mostly conservative. This is not exactly downtown San Francisco."[34] But even before the Dover ruling, CBS had paired Bush's advocacy of teaching the controversy with poll data showing 65 percent of Americans believed evolution and creationism should be taught side by side.[35]

Conciliation or Surrender?

A political disagreement suggests an issue that is conducive to a sort of joint-committee compromise. Political battles, especially at the national level, usu-ally are fights for the center. The two-party system tends toward moderation because third parties and radical elements usually do not need to be accom-modated. For issues where no "middle" is apparent, the two sides still may negotiate, which is made possible by the existence of only two sides and the very high prospective cost of appearing to be radically uncompromising in a centrist political culture in which moderation usually is a virtue. When media discover a political story, a natural approach is to provide balance in the coverage, giving voice to the process that tries to find a middle ground. Even an immoderate and uncompromising interest serves its cause well by appearing to be reasonable and simply looking for a place in the debate.

For example, Time magazine cast the Kansas story as a fight between "con-servatives" and "moderates," but reported that many scientists believed in both evolution and God. Another Time articled cited a Harris poll showing 55 percent of respondents said intelligent design ought to be taught alongside evolution. Other polls, Time said, showed 45 percent of people believing faith and Darwinism could not be reconciled. Time alluded to Dawkins, giving status to creationism by putting ID on the same plane as a leading scholar of evolutionary theory.[36] Articles in Time were fairly consistently critical of intelligent design, but dutifully gave space to critics of evolution. An essay in November 2005, "What Was God Thinking? Science Can't Tell," stated: "But as exciting as intelligent design is in theology, it is a boring idea in science. Science isn't about knowing the mind of God; it's about understanding nature and the reasons for things. . . . Intelligent design is a dead-end idea." A week later, the magazine noted, "Virtually no scientist agrees with i.d. proponents."[37] Calling creationism a "half-century campaign," Charles Krauthammer speculated that the nation simply may have been fed up with the ACLU "kicking crèches out of municipal Christmas displays." *Time* said ID was a "backlash" against evolution, a "new balance," but criticized intelligent design as an invasion of science in the same essay. Calling evolution "one of the most powerful and elegant theories in all of human science," *Time* acknowledged the frustrations of religious people. "How many times do we have to rerun the Scopes 'monkey trial?'" the reporter asked.[38] The president's entry into the controversy validated *Time*'s treatment of the story as politics. *Time* said his statement seemed "supremely fair minded: What could possibly be wrong with presenting more than one point of view on a topic that divides so many Americans? But to biologists, it smacked of faith-based science."[39] In spite of ample space to both sides, the political middle remained elusive.

Like *Time*, *Newsweek* treated the issue as a political story, and sought a middle ground in spite of being very tough on intelligent design at times— "biologists overwhelmingly dismiss it as nonsense." Bush was accused of "politicizing science" and of being a "willing tool" of creationists. *Newsweek* called intelligent design "a critique of evolution couched in the language of science," but quoted a critic of evolution as saying that by reducing everything to chemical reactions, "as Darwinians do . . . [evolution] is more ideology than science."[40]

Newsweek portrayed Dover as just another chapter in the Scopes saga, and quoted historian Edward Larson on the confusion over the word "theory," akin to "hunch" in its everyday use, but in science it means a systemic framework to explain empirical observations. However, a few paragraphs later, creationist Duane Gish was noted for fitting "every observation from paleontology, astronomy, and nuclear physics into a theory derived entirely from the Book of Genesis." Most scientists saw no issue, and a petition signed by 350 scientists dissenting from Darwinism was noted next to the membership of 120,000 for the American Association for the Advancement of Science.[41] The ratio was grossly disproportionate, but coverage was balanced, appearing to the nation as a two-party, partisan disagreement. Thus, President Bush's advocacy of teaching the controversy seemed reasonable. But even the president's own science advisor pointed out there is no real debate. A teacher in Kansas called ID "creationism in a cheap tuxedo."[42] *Newsweek*'s coverage reflected public confusion in that the magazine presented it as a polarizing issue in which one was either a creationist or an evolutionist, but insisted that a middle ground must exist.[43]

The national press fell prey to its own standards in accepting the political frame for intelligent design. Media gave the nation a story of political strife stripped of theological and scientific complexities and of history. In the press, finding fact was not so much a matter of verifiability as it was demonstrating impartiality. A reporter who gave voice to both sides in a controversy adhered to professional standards of fairness and fact. What was problematic in this case was "non-verifiable facts," the realm in which one might fit intelligent design. Reporters are absolved of malpractice and partisanship by presenting both sides of a controversy, presumably leaving the audience to decide who is telling the truth.[44] Creationists artfully manipulated the journalistic standard of fairness and made themselves appear part of a scientific discourse, enhancing their image as political moderates and as scientists.

However, it was not as though the press was helpless in publicizing bad science. The press had fallen victim to its own practices and standards, according to Chris Mooney and Matthew Nisbet, in a 2005 *Columbia Journalism Review* article. They took the national press to task for its coverage of the Dover trial. The abundance of causes for the problematic coverage included sending political reporters to cover the story. Not surprisingly, the reporters wrote political stories. This favored intelligent-design advocates:

> In strategy-driven political coverage, reporters typically tout the claims of competing political camps without comment or knowledgeable analysis, leaving readers to fend for themselves.
> . . . [S]uch "balance" is far from truly objective The pairing of competing claims plays directly into the hand of intelligent-design proponents who have cleverly argued that they're mounting a *scientific* attack on evolution rather than a religiously driven one, and who paint themselves as maverick outsiders warring against a dogmatic scientific establishment [emphasis in original].[45]

Opinion writers, too, were at fault, according to Mooney and Nisbet, because so many editorial page editors saw themselves as "enablers of debate within pluralistic communities—even over matters of science that are usually adjudicated in peer-reviewed journals." Mooney and Nisbet criticized the *New York Times* for abetting "a false sense of scientific controversy" when the paper ran an op-ed column by Michael Behe, who wrote a "scientific" critique of evolution.[46]

Television, Mooney and Nisbet wrote, was even worse in providing context, and "even the best TV news reporters may be hard-pressed to cover evolution thoroughly and accurately on a medium that relies so heavily upon images, sound bites, drama and conflict to keep audiences locked in." TV, political news, and opinion pages, they said, tended "to deemphasize the strong scientific case in favor of evolution and instead lend credence to the notion that a growing 'controversy' exists over evolutionary science. This approach may be politically convenient, but it is false." Given such coverage, it was no surprise that the general public was confused about evolution. Such coverage "only helps proponents present intelligent design as a contest between scientific theories rather than what it actually is—a sophisticated religious challenge to an overwhelming scientific consensus."[47]

Journalists have been slow to criticize intelligent design for another reason, and that is some recognition of the mainstream press' tendency to ignore or malign minorities. Journalism historian Roger Streitmatter cited this tendency as one of several ways in which the press has shaped American history. So any recognition inside the press of such an inclination may mean an attempt in contemporary media to err on the side of giving attention to all, even when the case of the minority appears to be so self-evidently poor, as with the creationists. Streitmatter cited numerous cases. The *New York Herald* called the 1848 women's rights meeting in Seneca Falls the "Women's Wrong Convention." The *New York World* denigrated Susan B. Anthony for her role in women's rights. In the 1930s Father Charles Coughlin used radio to attack Jews. More recently, Rush Limbaugh demeaned women, homosexuals, and people of color.[48] When it came to yet another group outside the mainstream, press sensitivity to that group may have been exaggerated by awareness of such episodes of maligning outside groups and causes.

Faith in Democracy

Just as Bryan had done in the 1920s, creationists in the twenty-first century used Jeffersonian ideals to impose a religious doctrine on fellow citizens. Bryan distorted egalitarianism to mean that all issues, including science, were a matter of the ballot box. Jefferson would not have agreed. Bryan's perversion of the ideal pervaded the creationist campaigns in Kansas and Dover. Bryan's faith in the people was grounded in sincere conviction on his part and a substantive career as a political reformer. In Dover, the creationists were sincere, proved adept at winning attention, but, like Bryan, did not understand science or its language. Bryan's primary concern was not science but changes in society and the decline of traditional values, which included religion. In the 1920s, evolution was among a number of ideas that traditionalists found offensive. In Dover, too, evolution threatened some Christians and was not the only offender of fundamentalist sensibilities. Like Bryan, his intellectual heirs in Dover clearly were out of their element when discussing science, especially on the witness stand. But those heirs also knew how to draw a crowd and win headlines. In the end, they also lost only in the courtroom, not in the court of public opinion, based on polls about belief in evolution.

The American tradition of antielitism/anti-intellectualism permeated both events. In the Scopes trial, the scientific experts for the defense could not testify. In Dover, they did, to the detriment of creationists. What may have been revealing about literalists' constancy over eight decades is the use of the word *theory*. In 1925, Bryan and others used it as a deprecating term, synonymous with a "guess." So it went in 2005. In public discourse, including the media, the word was commonly a synonym for "guess," without respect to scientific meaning, data, or evidence. Intelligent-design adherents meant it as a slight in reference to Darwin, figuring to devalue his theory with a popular audience. The elastic meaning of the word in popular parlance also meant that it could refer to nonscientific ideas. So in the public and media, intelligent design became a "theory," just like evolution. Deadline logic and the public's sound-bite attention span made intelligent design an alternative theory, even for those who did not like it. Such careless use of the word *theory* complemented the political nonpartisan inclination of the press. That was a creationist triumph.

The appeal to individual rights was another page out of Bryan, who as a reformer had consistently fought for individual rights and dignity in the face of overwhelming corporate and government power. In his heyday in the late nineteenth and early twentieth centuries, he had a legitimate complaint against railroads and trusts. He lost that legitimacy at the Scopes trial when he tried to make individual faith a measure of scientific validity. The same thing happened to Behe on the stand at Dover. He tried, under oath, to substitute personal faith for scientific fact. Like Bryan, he failed. In each case, though, the appeal to individual rights resonated with the culture. There is little evidence to suggest either man won many converts, nor did they lose any. Bryan's testimony, "final speech," and death meant only shifting tactics for fundamentalists in the 1920s. Behe continues to write, and polls continue to show most Americans support the idea of a sudden, perhaps young, creation.

The trial in Dover demonstrated that over the years since Scopes, creation-ists had successfully recast their literalism in a way that accommodated press and cultural values, gaining both attention and credibility for an idea that has no credibility in the scientific community. Creationism had moved from a faith-based argument to one that claimed to be empirically grounded and an argument not about religion in schools but about individual rights and fairness.

6

Into the Mainstream

The miscellaneous audience wants to listen to a man who
knows. How he knows is of no concern to them.

—Clarence Darrow

Dover's "Scopes 2" and Kansas's design debacle were signal events, showcasing phenomena that no longer represented a mere oddity in American culture. Though creationists lost in Dover and Kansas, the events reflected a movement that had crept from an intellectual backwater to the political mainstream.[1] Before 2005, the teaching of evolution already was designated marginal to failing in half of the states, as numerous legislatures and local school boards avoided, disclaimed, and renounced evolution. A 2001 report on teaching evolution showed a "passable C" average across all states, of which nineteen had less-than-passable standards. Of those "failing" states, large percentages of biology teachers believed creationism should be taught alongside evolution, and in some cases actually taught creationism in their classes. More states were deemed failing rather than excellent in teaching evolution.[2] Creationists held a Congressional briefing in 2000 and won a presidential endorsement in 2005. Gallup and Pew Research Center polls showed creationism's prevalence: According to a 2001 Gallup poll, 45 percent of Americans agreed that "God created human beings pretty much in their present form at one time within the last 10,000 years," which echoed results in its other polls on the topic since the early 1980s. A Pew study in August 2006 showed that 64 percent of Americans believed creationism should be taught alongside evolution.[3]

Creationists' patience, persistence, and campaigning had paid off. "Evolution or God" became part of the battle line that defined one's allegiance in the culture war. There was scant public criticism of those who rejected scientific evidence for Earth and life being millions of years old. There were some critics, but they were simply an opposing voice in a debate—the same news-story template in any number of national issues such as defense spending, national deficits, and illegal immigration. In this respect, creationists had triumphed because they had won recognition as the alternative voice. Intellectually, literalists remained in the 1920s, still fighting modernism and Darwinism. Politically, they were in the twenty-first century.

The Discovery Institute and *Darwin on Trial* were important markers in the creationist movement in the 1990s. The institute and the book reflected political sophistication and maturity, giving creationists a voice that sounded scientific and a theology that appeared grounded in both faith and logic. No longer a shrill minority casting aspersions on elitist universities or upon scientists whose research excluded faith, creationists now had the patina of intellectual credibility. It was not an accident. The *Wedge Document*, their map for winning public and political opinion, was working because by those first years of the twenty-first century, creationists had:

- Made belief in evolution (i.e., biblical literalism) a standard question in presidential elections.
- Produced numerous popular publications, many by people with a wide variety of credentials, some scientific, supporting intelligent design/creationism.
- Become a national issue by virtue of a new wave of activism in the U.S. Congress and state legislatures across the country.
- Reached out to ever-wider audiences and drawn more media attention with museums, conferences, and other public events.
- Raised millions of dollars and established institutes, fellowships, and publications to legitimize creationism as science.
- Created channels of communication and recruitment via new media.

Creationists opened a new, national front in May 2000 when they moved on the U.S. Congress. A congressional briefing, sponsored by the Discovery Institute, featured Phillip Johnson and Michael Behe, among others, in a three-hour informational session for congressmen and staffers. The event was titled "Scientific Evidence of Intelligent Design and Its Implications for Public Policy and Education." The congressional sponsors included Thomas Petri (Republican, Wisconsin), who introduced several of the speakers; Mark Souder (Republican, Indiana); Roscoe Bartlett (Republican, Maryland); and Sheila Jackson-Lee (Democrat, Texas). Charles Candy (Republican, Florida) provided the meeting room. The turnout was modest, about fifty people, only a dozen of them congressmen. At the time, however, Petri was in line to chair the House Committee on Education and the Workforce. He endorsed the institute's message. This was a new level of legitimacy and influence for creationists. The forays into state legislatures over the previous eighty years had kept the issue before the public and made it relevant to communities. The congressional briefing revealed greater ambitions, beyond states and local communities. Johnson and Behe reiterated Bryan's themes: that open-minded people were taking on a scientific elite, that creationism/intelligent design was scientific, that Darwin's theory was flawed science, and that evolution had banished God by replacing Him with a materialist philosophy. Speakers at the briefing reminded congressmen it was a matter of empowering people, who were sympathetic to creationism but misled by antireligion scientists. The briefing coincided with House and Senate consideration of legislation to revise federal education programs for kindergarten through high school, including math and science.[4]

Seven faculty members at Baylor University protested the event and sent Rep. Souder a letter denouncing intelligent design on the usual grounds: it was not science; there was no peer review of the intelligent-design research; it was not empirically grounded; it inevitably resorted to explaining nature as the will of the designer; the basic appeal was emotional, not scientific. Souder was indignant. His response, which he entered into the *Congressional Record*, stated that intelligent design had just as much data as "any other theory to determine whether an event is caused by natural or intelligent causes"; that "qualified scientists" supported intelligent design; that scientists and science education demonstrated an "ideological bias." Souder's letter, which was drafted by Phillip Johnson, focused on what he said was the Baylor scientists' motive: "to protect, by political means, a privileged philosophical viewpoint against a serious challenge." Worse yet, they treated "Darwinism as a straightforwardly scientific position" in spite of its "controversial philosophical premises."[5]

Rick Santorum, Republican senator from Pennsylvania at the time, may have come as close as anyone to making creationism a federal policy when, in 2001, he attempted to amend the No Child Left Behind Act. His amendment would have promoted teaching intelligent design in public schools by requiring schools to teach scientific controversies, specifically citing evolution as one such controversy. Johnson authored most of what came to be known as the Santorum Amendment, which was adopted 91–8 in the Senate but eventually removed from the final bill. It was relegated to a conference report on the legislation. Santorum persisted in his advocacy, citing academic freedom, "American pluralism," and elitism. "Anyone who expresses anything other than the dominant worldview is shunned and booted from the academy."[6]

Santorum wrote that intelligent design was in "contrast" to Darwin's theory: "intelligent design is a legitimate scientific theory that should be taught in science classes":

> Dissenting theories should not be repressed, but discussed openly. To do otherwise is to violate intellectual freedom. Such efforts at censorship abrogate critical thinking and will ultimately thwart scientific progress.
>
> Stifling freedom of discussion is wrong because it undermines the pursuit of truth and the presentation of different points of view, which should be the primary goal of education.[7]

Santorum said it was a matter of "intellectual freedom" and constitutional rights. "There is no reason to ignore or trivialize scientific issues involving controversial theories." He called his amendment "pro-education, pro-learning." Santorum softened his position in 2005, when he told an NPR interviewer that he would not go so far as President George W. Bush in allowing intelligent design to be taught in schools. "I'm not comfortable with intelligent design being taught in the science classroom. What we should be teaching are the problems and holes and I think there are legitimate problems and holes in the theory of evolution."[8] Santorum appeared to shift again later that year, with his resignation from the Thomas More Law Center, which had defended the Dover school board's intelligent-design policy. The resignation came only a few days after the federal judge handed down his decision against the Dover board in December 2005. The *Boston Globe* wondered whether Santorum's resignation, during a tough reelection campaign that he eventually lost, came after he noticed that the creationism advocates on Dover's school board lost their reelection bids the previous month.[9]

In spite of Santorum's qualified commitment, he affirmed his sympathy for the cause in a brief Foreword to the 2006 volume of paeans to Johnson, *Darwin's Nemesis: Phillip Johnson and the Intelligent Design Movement*. Santorum wrote that Johnson "provided extraordinary leadership for an extraordinary cause, namely, to rid science of false philosophy," which was "materialist reductionism, with its thoroughly unscientific denial of formal and final causes in nature and its repudiation of the first cause." Santorum tied the "decline of true science" to the decline in Western civilization: "There is much more for us to do, but working with Phil's colleagues at Seattle's Discovery Institute, we have begun the difficult fight for removing the stranglehold of philosophical materialism on textbook science."[10] Though Santorum's amendment was thwarted, the episode showed how creationism could become national education policy via the legislative process.

A National Campaign Issue

Creationism fully arrived as a presidential campaign issue in 1996, when Republican Patrick Buchanan embraced not just love of God and country but biblical literalism. Religion was nothing new to national politics and had been tied to various causes, such as reformers in the nineteenth century and progressives in Bryan's generation. In the mass-mediated era of post–World War II, Billy Graham became an unofficial consultant to a series of Republican presidents and aspirants to the office. In the '70s, Democrat Jimmy Carter made news with his public profession of faith. Conservative Christians were important to his presidential election in 1976, but many of those Christians found him too liberal, too moderate, and too inclusive. He lost to Ronald Reagan in 1980.[11]

In his years as California governor from 1967–1975, Reagan used his state board of education to promote teaching creationism in public schools. He supported the same idea in his 1980 presidential campaign. Not surprisingly, his secretary of education, William Bennett, supported the party line but was less strident, and he deferred to local communities to make the decision. It was a new plank in the Republican platform, though not yet a centerpiece, and the beginning of a string of candidates who adopted antievolutionism to appeal to conservative Christians. Reagan initiated the Republican embrace of a religious right that made teaching evolution a core concern.[12] During the 1980 campaign, Reagan's rationale for compromising the teaching of evolution was that it was only a "theory," one that had been challenged by other scientists: "But if it was going to be taught in the schools, then I think that also the biblical theory of creation, which is not a theory but the biblical story of creation, should also be taught."[13]

Reagan's approach was right out of the book of Bryan: equal time, let the people choose, and evolution was mere "theory." In another address, this time to the National Association of Evangelicals in 1983, Reagan assailed the intrusion of government into people's lives. In a very different context than evolution versus creationism, Reagan espoused the same egalitarianism and Jeffersonianism that guided his advocacy of creationism. In the speech, he condemned a law that allowed clinics to give birth control to teenage girls without notifying parents, "only one example of many attempts to water down traditional values and even abrogate the original terms of American democracy." He decried a judge's ruling in a Lubbock, Texas, case "that it was unconstitutional for a school district to give *equal treatment* [emphasis added] to religious and nonreligious student groups. The First Amendment never intended to require government to discriminate against religious speech."[14] Reagan's defense of religion in public places was akin to Bryan's conception of civic duty: putting God into the public sphere to promote a better society and citizenry. Reagan's appeal was populist in that he deferred to tradition, choice, and democracy.

However, explicitly embracing biblical literalism and rejecting evolution had not been part of a candidate's platform before 1996. In that campaign, Buchanan went beyond the conventional profession of faith and categorically denounced evolution, in part for its modernist spawn—i.e., liberal or left-wing causes. More specifically, he condemned evolution because it could not be reconciled with biblical literalism:

> Children should not be forced to believe the Bible, but I think that every child should know what is in the Old and the New Testament. . . . I believe the literal New Testament is literally the word of God, and I believe the Old Testament is the inspired word of God. . . . I think they (parents) have a right to insist that Godless evolution not be taught to their children or their children not be indoctrinated in it.[15]

The declaration jolted political commentators, who in a few cases seemed to have a hard time believing what they were hearing. A *New York Times* editorial puzzled that Buchanan's declaration "did not produce a major article or editorial proclaiming the candidate's views on evolution to be simple nonsense. . . . [W]hen a serious candidate for the highest office of the most powerful nation on earth holds such views you would think that this commentary would automatically become 'newsworthy.'" The *Times* pointed out that the need for presenting two sides existed only when there were two sides: "Nonsense masquerading as truth has been with us as long as records can date. But the increasingly blatant nature of the nonsense uttered with impunity in public discourse is chilling. Our democratic society is imperiled as much by this as any other single threat, regardless of whether the origins of the nonsense are religious fanaticism, simple ignorance or personal gain."[16]

The *Times* stamped the Scopes-trial imprimatur on Buchanan's position with the comment, "Mr. Buchanan played William Jennings Bryan with Sam Donaldson: 'Sam, you may believe you're descended from monkeys. . . . I think you're a creature of God.'"[17] Buchanan won only about 20 percent of the vote in the Republican primaries. In spite of the *Times*'s condemnation of his "nonsense," that 1996 campaign was the beginning of a faith test in presidential campaigns, particularly among Republicans. In the next presidential campaign, George W. Bush stated his support for teaching creationism alongside evolution. "I'd make it a goal to make sure that local folks got to make the decision as to whether or not they said creationism has been part of our history and whether or not people ought to be exposed to different theories as to how the world was formed." It was pure Bryanism. Bush did not need to say much on the subject in the 2004 campaign because he had made clear the importance of evangelical religion to his presidency.[18]

In a May 2007 forum, early in the presidential election season, three Republican candidates declared themselves antievolutionists. When asked who did not believe in evolution, those raising their hands were Arkansas Gov. Mike Huckabee, Rep. Tom Tancredo of Colorado, and Sen. Sam Brownback of Kansas. Huckabee later told the *New York Times* he believed in a "creative process," but, "If you want to believe that you and your family came from apes, I'll accept that." He added that children should not be "indoctrinated" in schools, and he would not support teaching creationism "as if it's the only thing that they should teach." Brownback later wrote, "I find there are too many complexities in the cell and wonders in the mind" to believe in evolution. He supported teaching the controversy. In a press release, Tancredo said, "Evolution explains changes in life. Creationism explains its origin."[19]

A 2007 Gallup poll affirmed that creation-evolution was an issue more for Republicans than Democrats. Among Republicans, 68 percent said they did not believe in evolution, compared to 40 percent of Democrats and 37 percent of independents. Those beliefs, Gallup said, also accompanied some apparent confusion about the issue because, "About a quarter of Americans say they believe both in evolution's explanation that humans evolved over millions of years and in the creationist explanation that humans were created as is about 10,000 years ago." Gallup noted the Republican debate in which Huckabee, Brownback, and Tancredo expressed their own disbelief in evolution. This, Gallup said, was "preaching to the choir." However much confusion may have been in the mix, the survey showed a large number of people embraced creationism. Two-thirds of respondents in an earlier poll said creationism was definitely or probably true.[20]

President Barack Obama was a disappointment for *Answers*, a publication of the Answers in Genesis ministry. Obama had stated he wanted to keep God out of public classrooms and politics. *Answers* said the "Creator's place in policy debates" was not likely to get a lot a attention. "The media and the politicians may attempt to sideline the Creator by simply ignoring Him or scorning Bible-believers. But the Creator still sits on His throne and weighs every nation in the balance."[21]

In the early skirmishes for the Republican presidential candidacy in 2012, almost all contenders hewed to the precedent set by George W. Bush a few years earlier—teach the controversy. Gov. Rick Perry of Texas qualified his objection to teaching evolution by calling it a theory with gaps. Before withdrawing from the campaign, Perry called evolution a "theory that's out there," and erroneously stated, "In Texas we teach both creationism and evolution." Rep. Michele Bachman of Minnesota said intelligent design should be taught alongside evolution, while she affirmed her belief in intelligent design. Gov. Sarah Palin of Alaska had faded as a contender by that point, but two years earlier suggested she had doubts about evolution, which in 2008 she believed should be taught alongside intelligent design in public schools. Antievolution had become, in the words of writer Chris Mooney, "a Republican phenomenon." It was such a mainstay in Republican politics that it became newsworthy when, in September 2011, former Utah Gov. Jon Huntsman wrote, "To be clear. I believe in evolution and trust scientists on global warming. Call me crazy."[22] Huntsman was out of the race before it really began, having withdrawn in December 2011.

By 2011, and the early days of the coming presidential campaign, the issue seemed to be a normal part of the political landscape. The mainstream press such as the *New York Times* no longer was awestruck by ignorance when a contender proclaimed the truth of a 6,000-year-old Earth, as the newspaper had with Buchanan in 1996. Evolution-creation was just another item in an issue checklist that separated Republicans and Democrats, as well as moderate or liberal Republicans from conservative Republicans.

Bad Science and Good Politics

It may not have been ignorance of science as much as it was the candidates' understanding of electoral reality: A large bloc of Americans adhered to young-Earth creationism and have consistently done so over the past quarter-century. A Gallup poll in June 2008 reported that 60 percent of Republicans said humans were created in their present form in the last 10,000 years, compared to 38 percent of Democrats. Gallup noted that, since 1982, there had been consistent results with 43 to 47 percent of people agreeing with the creationist view, compared to 9 to 14 percent adhering to a "purely secularist evolution perspective that humans evolved with no guidance by God." Those holding an alternative view, that humans evolved but God guided the process, amounted to 35 to 40 percent of Americans since 1982.[23] In May 2005, as creationism was coming to a boil in Kansas and Dover, Pennsylvania, another Gallup poll showed an incredible 76 percent of Americans would not be upset if creationism were taught in public schools. The same survey asked people if they would be upset if schools taught that human beings evolved from other species. Sixty-three percent said they would not. Those high percentages of people accommodating both perspectives are hard to explain except as indifference or accommodation. Gallup described it as a large number—45 percent—being "permissive" on the topic: "it wouldn't upset them if either creation or evolution were taught in the local schools." Those with a preference, however, were more likely to be in favor of creationism. That was 30 percent, versus 18 percent for evolution.[24] In another poll a few months later, Gallup reported, "People who adopt an orthodox creationist view are less likely to have thought a lot about the issue, but more likely to say it matters to them which is correct." The public confusion about or indifference to the issue was reflected in the finding that 58 percent said creationism was definitely or probably true, and 55 percent said evolution was definitely or probably true. Thus, each perspective had a majority.[25]

Gallup acknowledged, and its surveys showed, that acceptance of creationism did not exist in a vacuum. The correlates for belief in creationism included lower education, being 65 and older, regular church attendance, and identification with the Republican Party.[26] Biblical literalism correlated to views on other issues, including opposition to same-sex marriage and tougher immigration restrictions. Southerners were more likely to be literalists than people from any other part of the country. In the South, 41 percent said the Bible was to be taken literally, compared to 26 percent in the East, 31 percent in the Midwest, and 22 percent in the West.[27]

On Darwin's 200th birthday, February 9, 2009, only 55 percent of respondents to a Gallup poll could link Darwin to evolution. Though the poll offered no explanation for the knowledge gap, it may have been reflecting the same indifference or ignorance suggested in earlier polls that showed people holding conflicting opinions (being both literalists and believing in evolution) and that lower levels of education correlated to disbelief in evolution. The poll reported that only 39 percent of Americans believed in evolution, while 25 percent did not believe in evolution and 36 percent had no opinion.[28] The inability to link Darwin and evolution would not be the fault of creationist literature, which constantly linked the two. Instead, it may more likely be a lack of education or even some mis-education. However much confusion existed about evolution or even how much views about evolution versus creationism conflicted, 30 percent of people in May 2011 believed the Bible to be literally true.[29] Creationism may have been scientifically irrational, but politically it made good sense.

Presidential Endorsement

When President Bush endorsed "teaching the controversy" in August 2005, he was responding to and catalyzing an issue in the larger culture. In an interview in the White House with several reporters from Texas newspapers, Bush said he felt that "both sides ought to be properly taught" in order to help people "understand what the debate is about."[30] The Discovery Institute commended Bush for "defending free speech on evolution, and supporting the right of students to hear about different scientific views about evolution." Maintaining a moderate-sounding political tone, the DI said it opposed "mandating the teaching of intelligent design," but supported teaching criticisms of evolution. Teacher rights, it said, also were at issue so that educators might avoid "persecution or intimidation" for allowing students to "voluntarily discuss the scientific debate over intelligent design."[31] The institute portrayed Bush's stand as one for free speech, individual rights, and vindication for its own rebel-scientists. Stephen Meyer, head of the DI's Center for Science and Culture, said, "We interpret this as the president using his bully pulpit to support freedom of inquiry and free speech about the issue of biological origins. . . . It's extremely timely and welcome because so many scientists are experiencing recriminations for breaking with Darwinist orthodoxy."[32]

A *New York Times* news story said Bush "seemed to be reading from the playbook of the Discovery Institute." The *Times* said the DI's Center for Science and Culture "has emerged in recent months as the ideological and strategic backbone behind the eruption of skirmishes over science in school districts and state capitals across the country. Pushing a 'teach the controversy' approach to evolution, the institute has in many ways transformed the debate into an issue of academic freedom rather than a confrontation between biology and religion." The *Times* credited the Discovery Institute with a "politically savvy challenge to evolution . . ., propelling a fringe academic movement onto the front pages and putting Darwin's defenders firmly on the defensive. . . . Detractors dismiss Discovery as a fundamentalist front and intelligent design as a clever rhetorical detour around the 1987 Supreme Court ruling banning creationism from curriculums. But the institute's approach is more nuanced, scholarly and politically adept than its Bible-based predecessors in the century-long battle over biology."[33]

Only a few weeks after Bush made his comments, "The Evolution Wars" made the cover of *Time* Magazine, which, in the name of balanced coverage, gave ample space to the Discovery Institute. *Time* said that, on the surface, the president's remarks seemed "fair-minded," but such thinking smacked "of faith-based science" to biologists. In addition, his comment provocatively rekindled a "turf battle that goes all the way back to the Middle Ages," and did so at a time when many saw U.S. science being under both political and competitive assault. The article said that legislation challenging the teaching of evolution was pending or had been considered in twenty states. Noting the heated debates in Kansas and Pennsylvania, the article said Bush "has doubtless emboldened those who differ with Darwin and furthered one goal of that movement: he has taught all of us the controversy."[34]

Backlash

The mid-decade crest of creationism witnessed some backlash in less conventional mass media. Perhaps the surest indication of creationism's arrival as an issue in mainstream American culture was an episode of *The Simpsons* dedicated to the topic. In May 2006, less than five months after the Dover decision, the controversy erupted in cartoon Springfield, the Simpsons' hometown, when a minister asked the school to teach "alternative theories to evolution." Lisa Simpson, the daughter in the namesake family, objected. She recounted at home a test at school in which every answer was "God did it." In short order, Lisa organized a "truth seekers" club at school, was arrested for "teaching non-biblical science," and brought to trial. In court, an "ACLU appointed liberal" defended her, while the prosecution marched in a witness with a "PhD in truthology from Christian Tech." Lisa fared poorly in court until her mother, Marge, handed husband/father Homer a bottle of beer during the proceedings. He was unable to remove the twist-off cap, went into a simian rage as he banged the bottle, hooted and leaped about. Finally, the minister who instigated it all, and was on the witness stand during Homer's outburst, yelled, "Will you shut your yap you big monkey-faced gorilla!" Then, the minister had to concede, under questioning, that he could not rule out the possibility of Homer being descended from an ape. It ended up with Lisa and the minister on good terms, acknowledging one another's beliefs.[35] On its web site, Answers in Genesis dismissed *The Simpsons* because its moralizing was "usually politically liberal." Admittedly, the site said, one can't take a cartoon too seriously, but "it does show that the popular culture is acknowledging the creation/evolution debate to be a major part of today's culture wars."[36]

Judgment Day was a more serious indictment of events in Dover. The 2007 documentary included trial reenactments, interviews with participants, and even historical context. *Judgment Day* criticized intelligent design but gave time to its advocates, who included Philip Johnson. Michael Behe refused to be interviewed for the documentary, but reenactments included his testimony and that of Kenneth Miller, who testified for the plaintiffs. The documentary included Behe's admission during trial that, by his definition of science, astrology would be scientific. The film recounted Forrest's devastating testimony about the telling changes in *Of Pandas and People* manuscripts, in which *creation* was changed to *design proponents*. In some cases *creation* became *cdesign proponentists* [*sic*] when the search-and-replace function worked almost too well for a speedy revision after the 1987 *Aguillard* case, which outlawed teaching creationism in schools. A major theme of the documentary was the community division that the case provoked. In his decision, Jones pointed out that the people of Dover had been poorly served by board members who tried to inject religion into classrooms. After his decision, Jones said in *Judgment Day*, he had received death threats. Kitzmiller, one of the plaintiffs, said she received hate mail.[37]

Answers in Genesis said that the documentary was one-sided, did not portray intelligent-design advocates "in a good light" and presented evolution as "incontrovertible fact." In particular, AiG said, PBS was a tax-supported institution that was not balanced in its coverage. Adding to the offense, PBS issued a teacher's guide. "In other words, the audience for the documentary will not just be this week's PBS viewers—untold numbers of young people in America's schools will be shown a video that presents evolution as fact." Furthermore, the political climate had changed and was "not at all open to free inquiry," according to AiG.[38]

The Discovery Institute would not allow its scientists to be interviewed for the documentary, DI said, because, of the "known bias of PBS and NOVA against intelligent design" and "past experience" with misrepresentations, "especially through the editing process." Not surprisingly, the Discovery Institute condemned *Judgment Day* as a "lopsided, incomplete portrayal of the Dover trial based largely on Judge John E. Jones's poorly argued and scientifically inaccurate ruling." The news release alluded to PBS's 2001 *Evolution* series as evidence of anti–intelligent-design bias.[39]

Another documentary that incited the Discovery Institute's ire was *A Flock of Dodos*, a 2007 film that explored the creation-evolution fight via interviews with advocates for each side. As for the dodos, it was both sides. Filmmaker Randy Olson is a Harvard-trained former evolutionary biologist, who taught at the University of New Hampshire. In 1994, he turned to filmmaking. His basic premise in *Dodos* was that intelligent-design proponents may be scientifically suspect, but they were adept at public relations, well organized, well funded, and amiable. The evolutionists in the documentary—often shown in poker-game conversations at Harvard—were the obverse: scientifically sound, but dismissive of the opposition, unable to communicate with a lay audience, and oblivious to the success of intelligent design in the world beyond their card game. While the scientists played cards and derided creationists, the latter were in the thick of campaigning in Kansas. Olson sided with the evolutionists, but found the intelligent-design crowd to be "nice" people. In the context of the state hearing in Kansas in 1999 on teaching creationism, Olson said many scientists simply boycotted the hearings, and those who did show up were "obnoxious." "They were vying with each other to see who could humiliate the [state] school board the most." "Public-relations wise they were a disaster." An intelligent-design proponent and board member he interviewed was a contrast: "Connie [Morris] has charisma. She's funny. She's fun. She's likable. And there's no anger coming out of her. This was my first major note on how these two sides are communicating."[40]

Olson said at the end of the documentary that scientists may be the dodos because they were not able—like the dodos—to change with the environment. Intelligent design, by comparison, appeared to have the upper hand, in spite of being "a movement that has stalled at the intuition stage." Because the intuition was untestable, it was not science. Lacking data and testable hypotheses, its science was stalled but not its campaign.

In a May 2007 news release, the Discovery Institute said the documentary was neither accurate nor honest and included "false and potentially defamatory statements about Jonathan Wells and the Discovery Institute." Specifically, the release said, Olson misled viewers about Wells's antievolution book *Icons of Evolution*. The film accused Wells, an intelligent-design advocate, of being inaccurate in the book's assertion that early-twentieth-century embryologist Ernst Haeckel's drawings of embryos were reprinted frequently in modern biology textbooks. The drawings have since been discounted as evidence of evolution. "However, Olson's assertion that these embryo drawings have not been used in modern textbooks is absolutely false." Secondly, the release said, Olson was wrong about the Discovery Institute's budget, which it said had never reached $5 million, and was spent on many other topics than just intelligent design. Its Center for Science and Culture "is not primarily devoted to public relations, as the film insinuates."[41]

Answers in Genesis also disliked the film. AiG accused Olson of failing to differentiate between biblical creationism and intelligent design, which were "similar in some respects" but "really quite different." Creationists, AiG said, were much better at explaining "so-called 'poor design' in nature." The article did not elaborate on the point. AiG also claimed Olson erred in asserting that Haeckel's embryo drawings were not found in current textbooks. "He [Olson] comes across as an alarmist, for he believes that the IDers are becoming too successful, when in reality the movement has suffered setbacks in recent times (in the courts, with school boards, etc.)."[42]

In each case—*Simpsons, Judgment Day, Flock of Dodos*—the response from the creationist/intelligent-design side of the controversy was rapid and well considered for, and well targeted to, its constituency.

States' Nullification of Science

The dismal state of science education in the states may have been the most striking evidence for *Wedge Strategy*'s success. A 2000 report by the Thomas B. Fordham Foundation, *Good Science, Bad Science: Teaching Evolution in the States,* said nineteen states were doing a poor job of dealing with evolution in science standards. Among those states, the study's author gave Kansas special status—an "F minus"—because it was "unique in the extremity of its exclusion of evolution." The Scopes trial's home state of Tennessee maintained its notoriety by earning a failing grade. The report noted that nationally the controversy existed only in public schools, not in universities. It attributed this to K–12 education being subject to actions of legislators and school board members who often had little or no knowledge of the fields for which they dictated standards. In universities, by contrast, faculty who were expert in the fields were generally in control of curriculum and course content. *Good Science, Bad Science* cited evangelical Protestants as the "largest and most significant bloc that opposes the teaching of evolution in public schools." Pointing out that U.S. Supreme Court decisions had made teaching creation-science illegal, the report categorized state responses to creationists' pressure:

- avoiding the word evolution.
- simply ignoring evolution.
- some treatment of plant and animal evolution, but ignoring human evolution.
- deleting discussions that imply an old Earth or universe.
- the use of creationist "jargon."
- textbook disclaimers of evolution as "controversial" or "theory, not fact."[43]

These antievolution approaches were employed across the country. Many of the efforts were unsuccessful in diminishing evolution or in promoting creationism. But their frequency, persistence, and commonality of tactics were evidence of a groundswell of support for creationism and demonstrated the effectiveness of attacking the issue at the state level, which included not just legislatures but state boards of education and state curriculum committees. From 2001 to 2009, an average of ten antievolution bills were offered annually in state legislatures. The most popular approaches were allowing the teaching of "alternatives," which inevitably meant creationism (Alaska, Indiana, Michigan, New Mexico, Ohio); asserting that evolution was an "unproven theory," just a theory, or was theory and not fact (Alabama, Arkansas, Michigan, Nebraska, Oklahoma); and requiring the teaching of evidence against evolution (Arkansas, Georgia, Idaho, Ohio, Pennsylvania). In three states (Alabama, Illinois, Kentucky), "evolution" became "change over time." Only one state, Mississippi, simply excluded evolution from the state standards. In most cases, the proposals failed to pass a vote, never came up for consideration, or died in committee.[44] The level and pattern of activity suggested a coherence of organization and platform that was a far cry from the splintered denominationalism of Bryan's era.

Kansas was a leader among antievolution states at the time and an apparent inspiration to creationism activists when it excised evolution from the curriculum in 1999. Elsewhere, the Michigan House of Representatives in 2001 considered a bill that required equal time and referred to evolution as an unproven theory. In Kentucky, the education department in 1999 deleted "evolution" from state standards in favor of "change over time." Seven years later, Kentucky Gov. Ernie Fletcher, a Republican and Primitive Baptist, told the Kentucky Academy of Sciences that intelligent design should be taught because it was a "self-evident" truth to 90 percent of the population.[45] It was science because the majority said so. Bryan would have agreed.

Institutionalization

Creationist institutes, foundations, and fund-raising were flourishing by mid-decade. Such institutes and their accompanying network of fund-raising were part of a modern campaign structure that linked public institutions, communications networks, education, and funding sources. One study of creationist-institute finances found the total market for young-Earth creationist institutes more than doubled from 2003 to 2008. That analysis dropped one of the largest institutes, the Discovery Institute, because of DI's activities in areas other than creation research. Answers in Genesis was the largest creationist institute in terms of revenue, with its income more than doubling from 2003 to 2008 and accounting for 60 to 68 percent of what was deemed the total market for creationism. Donors were generous with AiG, with donations growing from $5.2 million to $7.7 million from 2001 to 2004.[46] By 2008, its income was more than $26 million, and in 2010 it was more than $22 million, of which almost $4.4 million was from admissions to its Creation Museum. Other sources of revenue included fund-raising events, radio, direct-mail appeals, product sales, and seminars. By comparison, income for the National Center for Science Education was $1.4 million in 2008.[47]

Before the arrival of Answers in Genesis, the Institute for Creation Research was the dominant voice among young-Earth creationist institutes. ICR is a sizable enterprise, with its income nearly doubling from 2005 to 2009, when it was $10.5 million.[48] Henry Morris, coauthor of *The Genesis Flood* (1960), founded ICR in 1970 in Dallas. It quickly became a central voice of the movement, and flood geology figured prominently in Morris's and ICR's creationism. The current chief executive office of the ICR is Henry Morris III, the eldest son of the founder. The younger Morris's graduate degrees include an MBA from Pepperdine University. ICR calls itself an educational institution, offering the MC Ed degree through its School of Biblical Apologetics. Media/publicity concerns are prominent in its mission statement: "ICR exists to conduct scientific research within the realms of origins and earth history, and then to educate the public both formally and informally through graduate and professional training programs, through conferences and seminars around the country, and through books, magazines, and media presentations." The ICR's books, films, periodicals and other media, including three radio programs, are available through about fifteen hundred outlets worldwide. The radio productions include a weekly fifteen-minute program *Science, Scripture and Salvation*, and a daily one-minute feature *Back to Genesis*.[49]

The Discovery Institute is the largest of the creationist institutes, though its work is broader than just promoting creationism. Its mission statement said nothing of evolution and focused instead on issues such as environment and the economy, transportation, government entitlement spending, and "religion and public life." The statement called DI "an inter-disciplinary community of scholars and policy advocates dedicated to the reinvigoration of traditional Western principles. . . . Discovery Institute has a special concern for the role that science and technology play in our culture and how they can advance free markets, illuminate public policy and support the theistic foundations of the West."[50]

Founded in 1990, in Seattle, DI originally focused on economics in the Northwest, but its emphasis shifted later to intelligent design, perhaps in part a result of substantial funding by Christian fundamentalists, and perhaps in part a matter of Director Bruce K. Chapman embracing the evolution-creation debate as the institute's signature issue. By 2008, its income climbed to $5.2 million. Its more than forty fellows included biologists, chemists, physicists, historians of science, and legal experts. In 2006, the institute claimed to have spent more than $4 million on research concerning intelligent design and evolution, and from 1996 to 2005 devoted almost 40 percent of its budget to intelligent-design research.[51]

The institute's impact on the debate has been substantial. The *New York Times* credited the Discovery Institute with nothing less than changing the debate over evolution from science versus religion to one of academic freedom.[52] The locus of the institute's intelligent-design activities is its Center for Science and Culture, which was created in 1996 with a grant of $750,000 over three years from Howard and Roberta Ahmanson, banking millionaires from Orange County, California, who had earlier given smaller amounts to intelligent-design research. By July 2008, the Ahmansons had provided more than a third of the center's $9.3 million since its beginning and underwritten about 25 percent of its $1.3 million annual operating budget. When the Discovery Institute opened an office in Washington, D.C., in 2004, it employed the same public relations firm that promoted the 1994 Contract with America, a Republican campaign document centered on smaller government and lower taxes. The institute's political stance was not without risk as it competed for resources in the political mainstream. DI's views cost it support in a few cases, as when a Templeton Foundation senior vice president in 2005 said the Discovery Institute was political: "[W]hat I see is much more focused on public policy, on public persuasion, on educational advocacy, and so forth."[53] Outside the intelligent-design research but within the realm of winning legitimacy, Discovery received in 2003 a $9.35 million grant, over ten years, from the Bill and Melinda Gates Foundation for the "Cascadia Project," which concerned developing transportation systems among partnerships in Washington, Oregon, and British Columbia.[54]

The American Scientific Affiliation (ASA) is one of the oldest creationist research institutes, founded in 1941 by evangelists with the broad mission of investigating areas related to Christian faith and science. In its first two decades, the ASA was an important forum for evangelicals wanting to debate evolution. It is not linked to a specific denomination, and includes Lutherans, Catholics, Methodists, Presbyterians, and Baptists. Focused more on appraising evolution than confronting evolution's proponents, the ASA is connected to Wheaton College, a Christian college in Illinois. The ASA is not strict young-Earth creationism, but more inclusive, including theistic evolution and old-Earth creationism, which became anathema to more conservative creationists, especially flood geologists.[55] The ASA's original goal was to promote the study of issues between science and religion, and that included evolution. The organization was not devoted to any single doctrine. That contributed to the eventual clash with young-Earth creationists who adhered to flood geology. The ASA's *Journal of the American Scientific Affiliation* ran divergent views. It stated in the forward of its January 1949 issues that it was dedicated to the truth of scripture and "elucidating the relationship of both the ideology and fruits of science."[56] Being nondoctrinaire may have cost it in terms of membership and revenue, and by 2009 its income was only $209,000.[57]

The ASA's young-Earth dissidents eventually split and formed the Creation Research Society (CRS), which dedicated itself to promoting literalism among young Christians.[58] The CRS, formally organized in 1963, began publication of *CRS Quarterly* in 1964. Its founding principles include the factual and historical truth of Genesis, including the Noachian flood. The CRS created the Van Andel Creation Research Center near Chino Valley, Arizona, to promote research into special creation. CRS income for 2008 was $286,000.[59]

On the National Agenda

In only a few years from the mid-1990s to about 2005, creationism won a prominent place on the roster of national issues. Pat Buchanan's advocacy of biblical literalism took people aback in 1996. It was one thing to court religious conservatives, as many candidates had done, but another to reject modern science. By the time George W. Bush endorsed creationism via "teach the controversy" in 2005, there were headlines, but not a lot of surprise. By the 2008 election, Republican contenders touted their antievolutionism as credentials for office. Institutes and foundations fueled the intelligent-design machinery, and reminded people that the Scopes trial—and its depiction in *Inherit the Wind*—was not merely history, but the defining moment in the Darwinists' war against people of faith. There was a lesson to be learned from Bryan's faith and tactics. The loss, creationists believed, had been a matter of failing to win public opinion, of surrendering politics to the secularists and ignoring the press in a mass-mediated age. Those things had been corrected.

7

Creationism's Web

In the Museum, On the Net, at the Movies

> Bryan: I do not think about things
> I do not think about.
> Darrow: Do you think about things
> you do think about?
> Bryan: Well, sometimes.
> —Darrow questions Bryan at Scopes trial

By the end of the twentieth century, the Scopes trial was no longer a humiliation for creationists but an indignation.[1] The problem, according to creationists, had been to make any compromise with literalism's naysayers. Creationists recast and remythologized Scopes as a lesson, their defining moment in the fight against evolution. They turned to a media and political arsenal—web sites, magazines, museums, institutes and lobbyists—to rewrite history as well as the rules of science.

The creationist showpiece, the $27 million Creation Museum in Petersburg, Kentucky, portrayed the Scopes trial as a seminal moment in modernist history. Though in agreement with the 1925 fundamentalists, twenty-first-century creationists found fault with Bryan's efforts. He had compromised. According to the museum's trial display, church leaders had it "the wrong way round. . . . The compromising theologians of that day refused to accept God's unchanging Word as their authority, and instead accepted the word of scientists—opinions that have since been revised and updated, and often discarded completely. The Scopes trial should be a 'big deal' to Christians—a reminder that a proper defense of the faith must begin with God's Word as the ultimate authority."[2]

For the Institute for Creation Research, the Scopes trial was nothing less than "a pivotal point in American history," a "debilitating defeat for Christians in the public eye." But it was a turning point because "(m)any Christian activists, legislators, and educators desire to reverse this ruling." It was the beginning of a "long campaign against Christianity's influence in America." By one ICR account, the ACLU "tricked" an unnamed lawyer into getting on the stand to defend the Bible, a lawyer who "himself had already compromised with long ages and local flood concepts, and couldn't give a consistent defense of either creation or biblical inerrancy." Another article on the ICR's web site was by that lawyer, who was Bryan. In the 1925 *Reader's Digest* article, he called evolution an "unproven hypothesis" that did not rise to theory because it lacked evidence. Some of Bryan's argument must have had a familiar ring for contemporary creationists: "Evolutionists also explain to us that light, beating on the skin, brought out the eye. . . . They also tell us that the leg is a development from a wart that accidentally appeared on the belly of a legless animal; and that we dream of falling because our ancestors fell out [of] trees 50,000 years ago." An "Editor's Addendum" to the web site republication said, "The article is of particular interest as a brief summary of some of the antievolution arguments of sixty-five years ago—arguments that, for the most part, are as cogent today as they were then." Bryan "was the most famous creationist of his day. Although he was not a scientist, his political eminence and oratorical ability impelled him into that role." The addendum warned of compromise, which earned Bryan "no sympathy, however, from either Darrow or the liberal press. The very fact that he believed God and the Bible was sufficient, in their view, to subject him to ridicule."[3]

In the twenty-first-century creationist narrative, Bryan basically had it right, but he erred in conceding that Earth might be older than a half-dozen millennia.

Creation Museum

Numerous small to moderate-size creation museums are scattered about the country, such as the "7 Wonders of Mount St. Helens—Creation Museum" in Washington; the "Creation Adventures Museum" in Arcadia, Florida; the "Dino Creation Museum" in Sacramento, California; and even a traveling "Creation Truth Ministries" museum headquartered in Alberta, Canada.[4] First among such enterprises is the Creation Museum. It was years in the making and was planned intensively. Answers in Genesis is the ministry behind the Creation Museum, and Ken Ham is the force behind the ministry. He is a former science teacher who moved to the United States from Australia to build his ministry. In 1994, he moved to northern Kentucky, choosing the area because nearly two-thirds of the U.S. population was within 650 miles of Cincinnati. By late 1999, Ham reported that he was directing a $5-million-a-year enterprise that included about 110 "creation clubs" across America. He was clear with the *New York Times* about the absolute position of the ministry's literalism and its role in the conflict:

> "I see a culture war hotting up in America between Christian morality and relative morality, which is really the difference between a creation-based philosophy and an evolution-based one," said Mr. Ham, who argues that if the word "day" in Genesis's account of creation is allowed to be taken metaphorically to encompass eons and to blur divinity's role, then the Bible is made fallible and morality reduced to human whim.
>
> Whether or not the United States is ready for such a culture war of words, Mr. Ham is. His group claims 140,000 on its mailing list and more than 3,000 visitors a day to its Web site.[5]

From the outset, Ham's idea was to have a more "layperson oriented" ministry than the Institute for Creation Research, from which Ham resigned in 1993 to establish Creation Science Ministries, which changed its name to Answers in Genesis (AiG) in order to more strongly reflect the ministry's devotion to the authority of all Scripture. Well before AiG, Ham's talent for growing an activist-political ministry was evident. At the ICR, he shifted the focus to nonscientists and aimed his media at a lay audience. In *The Lie: Evolution* (1987), Ham offered an accessible, engaging book on evolution as evil. The book used cartoons and anecdotes, sweeping generalizations, and a very simple theme—creation or evolution, salvation or damnation.[6] He turned the ICR from a marginal one struggling with finances, one whose speakers addressed crowds that often were no more than a few dozen and whose messages centered on trying to build a genuine scientific case for creationism via talks on such things as transitional fossils and biochemical processes. What did draw crowds were debates with evolutionists. Ham figured this out quickly and changed the message to a simple fire-and-brimstone attack on evolution. He began attracting crowds in the thousands, and the money started to flow.[7]

When Ham left ICR to establish his new ministry, he left behind ICR's concerns with scientific research. Instead, the energy and money went to media, always aimed at nonscientific audiences, especially young people. Within six months of launching AiG, Ham's radio program aired on 45 stations. AiG sponsored "teaching events," a newsletter, creation conferences, and even a scholarly sounding journal. It grew in only a few years into what is probably the world's most influential creationist ministry, and this via a media empire carrying a very simple message, astutely targeting a lay audience, redefining science, and just ignoring "experts" and mainstream scientists.[8]

The $27 million, 60,000-square-foot Creation Museum opened in May 2007, offering more than 150 exhibits. AiG said more than 100 credentialed media attended a ribbon cutting, including FOX-TV, CNN, ABC, and a number of leading newspapers. Within a few years, AiG said, its web site was drawing up to 94,000 visitors a day, and 3,000 to 4,000 people were visiting the museum every Saturday. By the end of its third year of operation, the museum had exceeded 700,000 visitors. In April 2010, the museum received its one-millionth visitor.[9]

The museum's opening generated attention across the country and even in Canada and London. The volume of news coverage was itself a triumph, an affirmation of importance and place on the nation's agenda. For the most part, reporters were impressed—even when critical—with the array of high-tech exhibits, including animatronic dinosaurs, a special-effects theater with vibrating seats and sprays of water, a planetarium, and the Dragon Hall bookstore and Noah's Ark Café. As for the thought and investment behind the museum, the designer was the same one who had helped create the "Jaws" and "King Kong" displays at Universal Studios in Florida.[10] What grabbed the most attention were displays about the coexistence of people and dinosaurs. This was AiG marketing savvy, not just because people like dinosaurs, but because dinosaurs are a special draw for children. The *Raleigh News & Observer* was one of the critical media: "[T]he Creation Museum will use people's fascination with dinosaurs as a draw." It was an enterprise the *News & Observer* quoted others as calling "pseudoscientific-nutty things" that portrayed "evolution as the path to ruin." Mainstream scientists, it said, "have dubbed it the Fred and Wilma Flintstone Museum."[11] The *New York Times* was just as impressed with the high-tech accoutrements and prehistoric children at play in Eden, "in which dinosaurs are still apparently as herbivorous as humans, and all are enjoying a little calm in the days after the fall." It was "sheer weirdness and daring."[12] The *Washington Post* cited others as calling the venture "nonsense," a "creationist Disneyland," and "widely discredited." But it was opening just a few weeks after three Republican presidential contenders had said they did not believe in evolution, the *Post* reminded readers.[13]

Protesters also were there for the opening. The *Chicago Sun Times* quoted a minister who said it showed, "Creationists are getting out of hand. . . . It's frightening because they're making inroads in public schools and in the presidential debate."[14] *USA Today* and *Newsweek* called the opening a success, as measured by thousands in attendance, media presence, and high-tech displays. Even demonstrators conceded the effective dazzle: "I'd give it a 4 for technology, 5 for propaganda. As for content, I'd give it a negative 5," said a professor from Case Western Reserve University in Cleveland, Ohio. *Newsweek* quoted the same person: "They have a right to build a museum. . . . What they don't have a right to do is to be openly fraudulent about science."[15]

From about fifty miles up Interstate 75, the *Dayton (Ohio) Daily News* found the technology impressive, but not the message:

> "This so-called Creation Museum is the institutionalization of a lie," said Lawrence Kraus, professor of physics and chemistry at Cleveland's Case Western Reserve University. "It's not religion. It's about scientific fraud."
>
> Sam Schloemer, a member of the Ohio Board of Education, said: "This systematic undermining of science education has dangerous consequences for our nation's future."[16]

NCSE's Eugenie Scott was part of a protest teleconference. She told the *Daily News*, "Teachers don't deserve a student coming into the classroom and saying, 'Gee, I went to a museum last summer and it told me you're teaching me a lie.'"[17] Another story quoted an Ohio State University professor of biology: "No qualified spokesperson or group for the scientific community recognizes any part of this as a museum." He called it part of a "transparent effort to recruit young people."[18]

Ham invoked the spirit of the Scopes when, at the dedication for the Creation Museum in 2000, he said the project would look to correct the damage done when Darrow put Bryan on the stand. More than a decade later, Ham still placed the trial and the center of the controversy. He faulted Bryan for not believing in six literal days of creation and accepting an old Earth:

> That's when Darrow knew he had won, because he had managed to get the Christian to admit, in front of a worldwide audience, that he couldn't defend the Bible's history (e.g., Cain's wife), and didn't take the Bible as written (the days of creation), and instead accepted the world's teaching (millions of years). Thus, Bryan (unwittingly) had undermined biblical authority and paved the way for secular philosophy to pervade the culture and education system.[19]

This, Ham wrote, was sadly the case among most Christians today who, "like Bryan, accepted the world's teaching and rejected the plain words of the Bible regarding history.[20] The museum's Scopes display called the trial the founding event for modern creationists.[21] *Answers*, a quarterly magazine published by Answers in Genesis, stated in a note to an article on Darwin:

> In the decades after Darwin published *On the Origin of Species*, evolution faced a mixed crowd of strong objectors, often religiously motivated. Their opinions on creation were diverse. . . . Some leaders in this crowd, such as William Jennings Bryan (1860–1925) of Scopes trial fame, even accepted evolution of plants and animals but not of humans. *It was out of this movement that the first modern creationists—those interested in developing a uniquely biblical and creationist understanding of the world and its history—first arose* [emphasis added].[22]

A *New York Times* account puzzled over the museum and its opening "at a time when conflicts between reason and faith, science and religion, keep erupting. Old debates about Darwin and biblical truth persist. Presidential candidates are asked about their faith in evolution. And 'Inherit the Wind' . . . is revived on Broadway." The article said that *Inherit the Wind*, upon second look, was "ill-tempered ridicule" that "has the air of agitprop." Recounting the story behind the play: "But the temptation to imitate the spirit of 'Inherit the Wind' is embraced by some who have angrily protested the Creation Museum, saying it presents an imminent danger, that it deserves the treatment and scorn the play gave to Dayton, or that the play's religious believers gave to its science teacher."[23] And, the *Times* noted, the play was not very good history. In another recollection of Scopes in the museum context, the *New York Times Magazine* said that at the time of the trial "creationists did not have a single credentialed supporter." But that had changed, and now the movement boasted any number of people with advanced degrees in science.[24]

Among broadcasters, CNN, like its print counterparts, found the technology impressive and the science dubious. CNN quoted Ham on exhibits showing dinosaurs and people coexisting, and the so-called evidence, but quoted others on the fact that "there's nothing scientific about the science at the Creation Museum." CNN even recalled the 1960s movie "One Million B.C. . . . where humans battled dinosaurs on prehistoric Earth. A new museum doesn't believe that story is fiction or fantasy, but a biblical fact." After quoting a Museum of Natural History official on "no scientific evidence" for a 6,000-year-old Earth, CNN cited Ham: "The purpose of the museum is really to give people information that's currently being censored from the public schools, from the secular universities, information they don't hear about that actually shows that evolution is not fact."[25] A consequence of such a story was to elevate, by association, Ham and the museum to the same level as the Museum of Natural History and its representative.

In a *Special Investigations* piece on faith and science that aired about a week before the museum opening, CNN put the Scopes trial at the beginning of the science-religion clash: "For much of the past century, science and religion have clashed here in America. The most fierce battle, the Scopes Monkey trial, when creationism squared off against evolution, and evolution won."[26] CNN found it was a free speech/censorship issue. Charmaine Yoest, of the Family Research Council, said, "Well, you know, mainstream science throughout history has been challenged by questions. And that's how we make advances in science, is being open to all different perspectives. . . . So we're really, really seeing an amazing censorship of anything that questions Darwinianism [*sic*]." She later was asked whether children should be taught ideas that have been disproven: "I'm not afraid of my kids knowing about any controversy that's out there as long as you put the evidence on the table, and consider what the debate is. That's what education is all about. It's having a vigorous debate."[27] The program's deference, as well as that of media, to the free-speech argument illustrated the effectiveness of the creationists' campaigns that had stressed the same point since Bryan.

Like other media, ABC praised the "high-tech sensory experience." But the network's coverage was tempered with skepticism of creation-science. The reporter said the museum was so sophisticated that scientists were concerned about its effectiveness in "giving children a distorted view of science." Another story focused on biblical literalism, with interviews of several literalists. One was an astrophysicist who used the rebel myth to bolster creationism's appeal: "Well, if you think about it, every major scientific discovery that's been made has gone against what the majority of scientists believe. . . . So the fact that [what] we're teaching goes against the majority really is irrelevant."[28]

Within a few months of the museum's opening, media were reporting on the 100,000th visitor to the museum.[29] Within three years, the museum had surpassed one million visitors. Flush with the museum's success and flooded with media attention, AiG decided to build an ark.

Ark Encounter

In December 2010, Answers in Genesis revealed plans for a $150 million theme park with a $24.5 million Ark Encounter museum. Kentucky Gov. Steve Beshear was on hand with AiG leadership for the announcement and coverage that included *ABC World News Tonight*, the *New York Times*, and *The Economist*.[30] AiG said it would be "an extremely powerful evangelistic tool," a "themed educational complex presenting historical accounts recreated from the Old Testament." The centerpiece would be a wooden ark, built according to biblical dimensions. The theme park would include a Tower of Babel, a first-century Middle-Eastern village, a petting zoo, and exhibits that depict the ten plagues and Moses's parting of the Red Sea. An animal trainer would present animal acts with a creationist message.

Scheduled to open in 2014, Ark Encounter is to be located on about 800 acres, forty-five minutes south of the Creation Museum, a little over an hour south of Cincinnati. Answers in Genesis expects 1.6 million visitors in the first year and anticipates 900 new jobs for the local economy and about 14,000 new jobs in area tourism-related businesses. Numerous AiG news releases emphasized economic impact, citing it as justification for a 75 percent property tax discount over three decades, as well as $40 million in sales-tax rebate from the state and $11 million in nearby interstate-exchange improvements.[31]

Gov. Beshear defended the tax incentives. He said the state could not discriminate against for-profit businesses based on subject matter, nor could the state deny the incentives based on religious grounds.[32] AiG and the governor apparently anticipated the constitutional issue. More than five months earlier, at a press conference announcing the project, the governor said the state would not "discriminate" based on subject matter of the park. AiG stated there was no constitutional problem, and cited "some of our most vocal critics" who conceded that no First Amendment issue was involved. Employment, AiG said, was the primary issue.[33] AiG told its media critics, which it said included the *New York Times*, that whatever the message, there was no First Amendment issue. Bolstering that argument was the fact that the AiG would not be the owner of the project. That would be Ark Encounter LLC, which was for-profit and would receive the tax rebates. AiG would be the designer and operator of the ark and would be the only member of the LLC.[34] In fact, AiG and Ham countered, the state was already supporting "the religion of Secular Humanism to the tune of billions of dollars per year through tax-supported public schools/universities, science museums, zoos, PBS-TV, etc. Secular Humanism is an actual religion that denies the one true God and worships self or science (or other idols)."[35] Ham charged that AiG and the Ark Encounter were victims of "viewpoint discrimination" by those who did not "believe in the freedom of religion that most Americans seek."[36] The underdog theme was as old as the Scopes trial for creationists. AiG used it again in the Ark Encounter: "With all the media attention (both mocking and embracing the project), one might imagine what was said about Noah when he began building the biblical Ark."[37]

Before physical work ever started on the Ark Encounter, it was a resounding success in terms of publicity, just as the Creation Museum had been. "Media outlets and bloggers worldwide have been reporting on our plan to partner with Ark Encounter LLC to build a Noah's Ark themed attraction," AiG said. The ministry found the media coverage encouraging—for both the Creation Museum and the Ark Encounter. A press conference concerning the latter provoked *New York Times* coverage and "Google alerts . . . almost every 5–10 minutes from all around the country (and world) newspapers, web sites, and blogs . . . commenting on the Ark announcement," as well as front-page stories in the *Louisville Courier-Journal* and the *Cincinnati Enquirer*. It was even covered internationally, AiG reported, as the ministry awaited a story from the *Times of London*.[38] Creationists' use of the internet complemented the success of the museum and the ark. AiG bypassed the skeptical or biased—by creationists' measure—mainstream media. Creationists countered the "establishment," whether media or science, via the web.

Answers on the Web

The creationist institutes were ready for the president's blessing in 2005. A day after George W. Bush endorsed teaching the controversy, the Discovery Institute's Center for Science and Culture issued a news release praising the president for his support of "free speech." DI stressed its support for teaching "scientific criticisms of Darwin's theory" and for teacher rights. The institute quoted Bush that it was "not the federal government's role to tell states and local boards of education what they should teach in the classroom" and that "scientific critiques of any theory should be a normal part of the science curriculum."[39] Three days after Bush's remarks, Answers in Genesis recounted the reports from various media and DI's stance but found Bush had not gone far enough:

> While design arguments in the Intelligent Design movement may seem very appealing at first, the central problem with the ID movement . . . is that it divorces the Creator from creation. The Creator cannot be separated from creation; they reflect on each other.
>
> The only real hope for rebuilding the broken foundation of our once-Christian nation is to return to the authority of God's Word, beginning with Genesis 1–11.[40]

That same day, Ham took the opportunity of the president's comment to ponder the gathering storms in Pennsylvania and Kansas. He saw the events as part of the ongoing war, and said the "creationist movement is having a large and dramatic effect on American culture. . . . There is a groundswell of growing attention to this topic across the nation. Why else would some of the evolutionists be so militantly concerned about what's going on? I think it's because of the effectiveness of what's happening."[41]

Answers in Genesis's multitude of web sites offered hundreds of videos, books, articles, and photographs ranging from children's Bible lessons to select PhDs affirming the validity of young-Earth creationism. Generally, the web sites were easily navigable, visually appealing, well written, and intellectually accessible in that they sound scientific and logical. The last point may cause a few scientific "harrumphs," but impressing mainstream scientists was not the point of the sites. AiG launched its web site in 1995 and by 2011 claimed about 25,000 visits daily.[42]

Creationists used the web, in part, to correct what they saw as the historical injustice, and public perception, of the Scopes trial. The myriad AiG web sites that addressed the Scopes trial reflected great concern about that piece of American history, especially *Inherit the Wind*. AiG bemoaned the stage version (1955) because it was, according to AiG, so often a choice for secondary-school stage productions. The ministry lamented the historical inaccuracies of *Inherit the Wind*, which, if one wanted a trial history, was a valid concern. AiG's main web site declared the ministry's commitment to fighting "superstition" and science "cults," which AiG said began with the Scopes trial. *Inherit the Wind* "savagely distorts the historical reality of the 1925 Scopes 'monkey trial.'" The movie was an "epic 'light vs darkness' myth—evolutionary science is 'good' and creationism is 'evil.'" The article stated that where "Darwin-rejecting churches are the weakest, there is the greatest flourishing of cults, occult activity and various forms of superstition." Cults included evolutionism, which was deemed a pseudoscience.[43]

AiG never mentioned the historical fact that the real target of *Inherit the Wind* was 1950s McCarthyism. The decades-old Scopes trial was merely a vehicle for comment on the intolerance and bigotry of Sen. Joseph McCarthy in his relentless search for communists in America, and his willingness to ruin lives and stain reputations in order to promote his own political career. However, in light of the tremendous popularity of *Inherit the Wind* and its large place in film history, its impact on popular impressions is a valid concern for religious fundamentalists. The film was an excellent foil for creationists, and multiple AiG web sites cited the film to demonstrate, among other things, that religious people

- are intentionally misrepresented in order to justify discrimination against their ideas, which means teaching creationism in public schools.

- are rebels to convention, and are outcasts from mainstream science and media, each with their own liberal and/or humanist agendas.
- are victims of a larger, secular conspiracy as evidenced by the continued embrace of a skewed history of the trial. On that point: "The evidence suggests that the inaccuracies in the play and film *Inherit the Wind* are *substantive, intentional and systemic* [emphasis added]. Christians, and particularly William Jennings Bryan, are consistently lampooned throughout the play, while skeptics and agnostics are portrayed as intelligent, kindly, and even heroic."[44]

The concern about the influence of *Inherit the Wind* was well placed because "Scopes 2" recurs with great regularity across the country. Even when cases don't go to trial, Scopes is the near-reflexive context provided in news media. Since *Inherit the Wind* is a familiar, if flawed, historical reference for so many people, it would be important to influence impressions about the actual event. The same web sites that sought to set the historical record straight often did so with a format that gave "myth" and "fact" of the trial alongside one another, such as the fictional torch-bearing mobs of the movie versus the historical reality of amiable town folk in Dayton, or the collapse of madly ranting Bryan at movie's close versus the fact that he died in his sleep a few days after the trial, or the jailing and physical threats against the Scopes character versus the fact that the real Scopes never was in jail and never was physically threatened.[45]

At times, AiG gave great power of suasion to Scopes, as in an article that stated, "Today, secular humanists are using the Scopes trial as a weapon to attack Christianity as a religion of blind faith—one that is unscientific and irrational."[46] AiG assailed media for relying on stereotypes from *Inherit the Wind*, for not providing balanced coverage (a charge made sincerely and without apparent irony, considering the leeway given to biblical literalists in a supposed science debate), and for failing to acknowledge the impact of young-Earth creationism in American culture. With the attention to Scopes on the creationist web sites, Bryan was inevitably prominent. The web sites invoked his antielitism and anti-intellectualism via attacks on "conventional" science and education, and they reiterated his themes of fundamentalists as the underdogs, the persecuted, the rebels.

The AiG web sites carried an old message in an effective fashion—a culture war of two, and only two, sides with diametrically opposed world views. The messages simplified the issue and the choices. Such an approach facilitated the bundling of issues, such as abortion, limited government, immigration, and teaching creationism in schools. In addition to a rapid response, the web platform provided the opportunity for unlimited content, which would be important not only in maintaining its existing network of sympathizers but could be especially helpful for potential recruits to the cause, supplying an endless stream of information to banish doubt and bolster creationism. When PBS offered several scientific series on evolution and human origins, the Discovery Institute and Answers in Genesis used the web to quickly provide allies and sympathizers with responses and pedagogical tools for combating the threat.

Correcting *Evolution*

In September 2001, PBS ran the seven-part series *Evolution*, which looked not only at the theory but also its cultural impact, from its origins with Darwin in Victorian England to the religious turmoil in contemporary America. In an apparent attempt to be balanced and inclusive, PBS devoted an entire episode to religion and evolution. The Discovery Institute's web pages erupted with ire over the show's treatment of science, evolution, and religion, and skewered the series for shortcomings, misrepresentations, omissions, and other perversions of truth. Some of DI's criticisms even preceded the airing of the first episode and used authoritative people with advanced degrees. In eight hours of video, PBS approached the subject historically, scientifically, and culturally, including the initial shock of Darwin's theory for scientists, theologians, and the general public; extinctions; adaptation and survival of the fittest; sexual selection; and religion. The concluding episode, "What about God?," opened with Ken Ham declaring, "It's a war." He stated that, as for creation, a "day" meant a 24-hour day, not an age. A Genesis day was 24 hours. Nothing else. Ham dismissed evidence for an ancient Earth and questions about how fossils came about in only 6,000 years. He stated that Noah's ark was a "real boat." The *Bible*, he said, should be taught in schools. NCSE director Scott appeared later in the episode, explaining what is and what is not science, the importance of testable evidence and excluding supernatural explanation.

The series dealt forthrightly with the existence of two mutually exclusive world views. The first episode gave creationists ample room. They insisted that science was not the issue and that the controversy was about freedom of choice and expression. In that episode, "Darwin's Dangerous Idea," Lafayette, Indiana, was a case study of the creation-evolution conflict. In interviews with students, it became clear that the creationist campaign had prevailed in at least one respect: the issue was not religion or science, but individual choice. Several students insisted that they should be allowed to decide whether special creation was science. They saw it as a matter of individual freedom. As to whether or not creationism was science, it did not come up among the students.[47]

The Discovery Institute responded to *Evolution* with *Getting the Facts Straight: A Viewer's Guide to PBS's Evolution*, a 150-page, online guide (also available in hard copy for $7.95 from DI's web catalog) that critiqued each of the episodes and provided different "Activities" to accompany each episode. Episode seven, "What about God?" inspired the most activities, with three: "The Scopes Trial in Fact and Fiction," "Who Are Darwin's Critics Now?" and "Teaching the Controversy: What's Legal?" The *Guide* accused PBS of producing a "one-sided piece of advocacy, unworthy of a publicly funded broadcast network."[48] DI made the latter point several times and may have given the alleged transgression special attention because that cast the program as political, an environment where creationists had competed successfully with evolutionists. In charging that the series failed to meet standards of fairness, the *Guide* stressed the political issue and, therefore, the obligation of a publicly funded entity to provide both sides:

> Clearly, one purpose of *Evolution* is to influence Congress and school boards and to *promote political action* [emphasis added] regarding how evolution is taught in public schools. . . . Imagine, for a moment, that PBS created a seven-part series on abortion . . . to influence national and local government officials regarding abortion legislation. . . .
>
> In summary, the PBS *Evolution* series distorts the scientific evidence, omits scientific objections to Darwin's theory, mischaracterizes scientific critics of Darwinism, promotes a biased view of religion, and takes *a partisan position in a controversial political debate* [emphasis added]. By doing this, PBS has forsaken objectivity, violated journalist ethics, and betrayed the public trust.[49]

The *Guide*'s authors seemed particularly nettled by the use of Scott as a spokesperson for the project. Her inclusion made the "political agenda behind *Evolution* . . . more explicit." The *Guide* accurately identified her as director of the NCSE, which it said was "a single-issue group that takes only one side in the *political* debate [emphasis added]."[50]

The *Guide*'s classroom antidotes to *Evolution* included "learning objectives," directions for student activity and discussion, and source citations. Activity 5, for example, accompanied the *Guide*'s chapter four, which was a detailed assault on Episode 4, "The Evolutionary Arms Race." The activity provided a select overview of the PBS material. In this case, the brief six paragraphs noted the importance attached in the episode to empirical research and being able to differentiate between "well-supported claims and conjectural ones—those that are founded on insufficient information and reasoning." The overview recounted Stephen Gould's criticisms of a "speculative style of argument" that had hampered evolutionary biology. "In other words," the *Guide* said, "what sometimes happens is that biologists will gather information about an animal's current traits and environment—then go beyond the data by inventing a scenario to explain how things got the way they are." The activity was worded in a way that could be read to mean that, in the PBS episode, Gould criticized evolutionary-biology research. That was not the case. Gould appeared in the episode, but his comment on the "speculative style of argument" was pulled from an Introduction he wrote to a 1980 novel, *Dance of the Tiger*. The novel is about prehistoric humanity; Gould praised the work for accurate scientific detail and said that such literature was a more appropriate place for speculation about the Neanderthal–Cro-Magnon issue than professional science research literature. A closer reading of Gould's remark and less selective choice of his Introduction revealed an idea that would appall creationists—purposelessness:

> Evolutionary biology has been severely hampered by a speculative style of argument that records anatomy and ecology and then tries to construct historical or adaptive explanations for why this bone looked like that or why this creature lived here. These speculations have been charitably called "scenarios"; they are often more contemptuously, and rightly, labeled "stories" (or *"just-so stories"* if they rely on the fallacious assumption that everything exists for a purpose) [emphasis added].[51]

So Gould's apparent criticism of evolutionary biology actually was a condemnation of what lies at the heart of creationism: purpose and order exist, and it is the role of science to reveal them. Instead, the Discovery Institute's activity explained the remark as meaning that biologists too often gather data "but then go beyond the data by inventing a scenario to explain how things got the way they are," i.e., by invoking evolutionary theory.[52] It was a very deft mining of Gould's voluminous writings on evolution.

In Activity 5, two of the learning objectives were making students aware that "even scientists may present conjectural claims as factual" and identifying "conjectural historical claims made in Episode Four." The "Directions" segment of the activity started with an eye toward spotting the conjecture on the part of scientists—"a conclusion or opinion that is based upon insufficient evidence." The classroom exercise asked students how to find out how fast a cheetah could run. A good answer, according to the lesson plan, would be clocking its speeds while it was chasing prey. The next scenario concerned those same properly measured speeds, but with the claim that "cheetahs got their great speed due to environmental pressures that forced them to run fast to catch game and survive. Is this a well-supported claim or conjecture?" The correct answer, according to the *Guide*, is "conjecture," because the change in speeds was measured at only one point in time and "cannot tell us whether the speed of the cheetahs has increased over several generations." The second class session devoted to Episode Four had students discuss the parts of the episode they found conjectural and why.[53]

The Scopes trial was prominent in the DI's *Guide*. In Episode Seven, PBS devoted less than a minute to the trial. The episode said "Bryan prevailed," and the trial resulted in a chilling effect on the teaching of evolution. The *Guide* devoted an entire activity (there were a total of eight for the episode) to the Scopes trial. The focus was not on the depiction of the event in the PBS series. Instead, it was an opportunity to "examine differences between the real Scopes trial and the fictional portrayal of the trial in *Inherit the Wind*," which presented the trial as a "stark showdown between defenders of free speech and religious fundamentalists who want to censor science teachers with whom they disagree." The overview of the activity correctly pointed out that *Inherit the Wind* was not good history. The objectives of the activity were to understand the differences between the historical trial and the dramatic version of the trial, and to help students "be aware of how their perceptions of historical reality may be shaped by films and television." To those ends, the *Guide* recommended viewing *Inherit the Wind* and reading historical accounts. A major problem with *Evolution*'s Episode Seven, according to the *Guide*, was the poor representation of religion:

> So out of the vast spectrum of the world's religious beliefs, *Evolution* gives voice only to biblical literalists—whom it dismisses as uneducated and doctrinaire—and to the small minority of Christians who subscribe to Darwin's theory. The series completely ignores the hundreds of millions of other Christians—not to mention Muslims, Hindus, and orthodox Jews—who reject the Darwinian doctrine that all living things—including us—are undesigned [*sic*] results of undirected natural processes.[54]

Activity 8 went to the heart of the issue: "Teaching the Controversy: What's Legal?" There was no suggested *Evolution* episode to accompany the activity, which was aimed at getting students "to dig deeper on the constitutional issues related to teaching origins in public-school classrooms." Among the objectives were identification of scientific criticisms of Darwin, describing core concepts of intelligent design, and assessing the First Amendment issues surrounding teaching either Darwinism or intelligent design in public schools. In the directions for the activity, a fictional, not-very-imaginatively named "John Spokes" wanted to make some changes in the way he taught evolution. He wanted to "correct blatant factual" errors that overstated Darwin's case in the textbook, tell students about new evidence that was not mentioned in the text, discuss the parts of evolution that "remain controversial" among scientists, and tell students that "a growing minority of scientists do see evidence of real . . . design in biological systems." It pointed students to law review articles, a U.S. Supreme Court case (*Edwards vs. Aguillard*), and a book published by the Foundation for Thought and Ethics, which earlier published *Of Pandas and People*.[55]

In all, the *Guide* was more than a rejoinder to *Evolution*. It provided ready-made criticisms, supplemental readings, classroom exercises—all in the name of free speech, critical thinking, individual liberty, public interest, true science, and theological liberalism. The *Guide* sounded like a scientific answer to the PBS series, and made the task very simple for creationism-inclined teachers, or teachers under pressure to accommodate creationism while being required by state standards to teach evolution and science.

In 2009, another PBS series, *Becoming Human*, provoked an AiG attack that was reminiscent of Bryan's assault on Darwinism: it was bad science, and it was not approved by the people. On the first count, AiG said, "Unfortunately, for those interested in the supposed jump from these ape-men to humans, 'the fossil record is virtually silent,' the program says. . . . Apparently creationists find the gaps far more problematic for ape-man evolution than do evolutionists (surprise, surprise)." On the second count, AiG said, "It bears repeating . . . : it is the U.S. taxpayer who helps fund expensive TV programs like this. . . . [T]hus, the 44% of Americans who believe that 'God created humans pretty much in their present form either exactly as the Bible describes it or within the last 10,000 years' will be dismissed." One comment paralleled Bryan's criticism that Darwin littered his work with "words implying uncertainty, 'obscure glance,' 'apparently,' 'resembling,' 'must have been,' 'slight degree,' and 'conceive.'"[56] AiG stated, "As we expected . . ., speculation runs rampant in this first segment on human origins. Words such as *may*, *somehow*, *believes*, *perhaps* and also phrases like 'some scientists still question . . .,' 'the fossil record is virtually silent . . .,' and 'debates rage,' all show just how weak the case really is for ape-like ancestry of man" [emphasis in original].[57]

Alleged Truth

The movie *Alleged* recast the Scopes-trial myth. The film's plot was a rejoinder to *Inherit the Wind*'s themes—small-town intolerance and ignorance, Bryan the fool, the wise and tolerant Darrow, and the triumph of reason. *Alleged* was the debut release in November 2011 of Slingshot Pictures, a company that focuses on faith-based titles. The screenwriter, Fred Foote, said it was his intent to rewrite history, specifically *Inherit the Wind*. "It's a great play, but when I read the actual trial transcript, I thought not only is it inaccurate, it's completely the opposite of what actually happened. . . . In my opinion, 'Inherit the Wind' is almost a photo negative of the real story." Instead, *Alleged* depicted Darwinism and evolution as the tools of the pseudoscientific eugenics movement, which was in its ascendency in the 1920s.[58] Part of the rationale behind eugenics was radical "social Darwinism" that advocated not just economic survival of the fittest, but improving the human race by promoting better breeding. The more radical eugenicists advocated keeping "less fit" individuals from propagating, usually by sterilization. Backed by family money, Foote set out to correct history and *Inherit the Wind*.[59]

Alleged wrapped the science-creationism conflict in a romance between Charles and Rose, in 1925 Dayton, Tennessee. Charles is a young reporter at the "Dayton Herald" and is led astray when he becomes enamored of Mencken's journalism, adopting the celebrity reporter's cynicism, cigar smoking, and drinking. Charles veers from the path of small-town virtue, but he manages to salvage himself and his romance by the end of the movie. The lovers even agree that Mencken cannot be trusted, especially in light of his support for eugenics and his revelation to Rose that she is of "questionable" lineage—a "feeble-minded sister whose father was a darkie," a mother who committed suicide, and an uncle in St. Louis who likes little boys. "That makes you, you," he tells her. All the while, the institution where her sister is housed is preparing to sterilize her, whether Rose approves or not. In the end, Charles and Rose save the girl and take her with them.

In *Alleged*, Scopes is convicted. But several trailers make sure the audience gets it right this time. Among other things, those trailers point out the following:

- "Nebraska Man . . . touted by America's leading paleontologist as 'irrefutable evidence' for human evolution [and alluded to in the movie] on the basis of a single tooth. Shortly after the trial, other scientists determined the tooth to have belonged to an extinct pig."
- Hunter's Civic Biology, from which Scopes taught, included "vicious statements regarding racial minorities, the poor, the disabled."
- The U.S. Supreme Court in 1927 upheld a law allowing 60,000 Americans to be involuntarily sterilized.

- Bryan College opened in Dayton in 1930.
- From Scopes: "There is more intolerance in higher education than in all the mountains of Tennessee."

The movie portrayed Mencken as acerbic, rude, thoughtless, and cynical, all of which have some historical justification. The fictional Mencken tells Charles, "Evolution is progress. . . . Don't get suckered in by their religion. . . . Science is reality. . . . Religion is fairy tales." The Mencken character capped his short screed with the statement that one "can't mix the two," as he steals an apple—of all things—from a street vendor.

The remake of *Inherit the Wind* did pretty much what it set out to do in terms of reinterpreting the trial: Liberals and scientists are intolerant. Progress may be found in big cities, but virtue resides in small towns. The evils of racism and eugenics are products of modernist thinking, a point also trumpeted in creationist publications and prominent in the Creation Museum.

Darwin's Immoral Science

Just as Bryan equated Darwinism with the nightmare of World War I, modern-day creationists credit Darwinism with World War II and the Holocaust. In his indictment of Darwin, Bryan did not distinguish among "evolution," "natural selection," "Darwinism" and "social Darwinism." It would have complicated the issue and the message, and complexity was not politically expedient. The Creation Museum showed more finesse: In its "Darwin Room," the museum distinguished between evolution and natural selection, stating, "Natural selection is not evolution." Elsewhere in the museum, however, the displays did not distinguish between Darwinism and social Darwinism. There is good reason politically not to make the distinction. A prominent, consistent objection to Darwin in the Creation Museum and in various AiG web sites and literature is the relationship among militarism, war, racism, and "materialist philosophy"—Darwinism, in other words. The museum's Scopes display video began with Bryan and his views of World War I, including his condemnation of social Darwinism, which Bryan linked to the rise of German militarism. The Creation Museum adopted Bryan's argument via displays that tied Nazi racism, eugenics in America, and even modern urban blight to Darwin. The museum's use of Darwinism as the nemesis of civilized society was true to Bryan's perspective in several respects. As noted, Bryan did not bother to distinguish among Darwinism, evolution, and social Darwinism. His real objection was to the latter and its implications for people and society. AiG also was indifferent to the distinctions but equated them with ignoring the truth of Scripture. In one display, "Modern World Abandons the Bible," the pseudourban area—the "Graffiti Alley" display—was replete with dilapidated homes in a filthy environment. It was an implicit contrast to the "garden" of rural America.[60]

The museum replicated Bryan's argument with Darwinism in another fashion. Bryan opposed the idea of "survival of the fittest," especially in reference to people and society. The museum lambasted survival of the fittest. Both the museum and Bryan acknowledged only one-half of the Darwinian equation. In *Origin*, Darwin wrote at length about the role of cooperation, as well as competition, in an organism's survival. Creationist literature rarely recognizes the cooperation part of Darwinism, perhaps because doing so would make more problematic the linking of Darwin to racism, militarism, and eugenics. *Answers* magazine stated: "Darwin believed creatures were at war with one another, competing for limited resources. In their struggle to survive, the stronger, faster, and more cunning survived and thrived, while the rest died."[61] Another museum display, "Who's your brother?" pointed out that, according to human reason, racism, and genocide are under the rubric of "biological arguments for racism." But, "according to God's word, . . . we're all one race—one blood."[62]

Creationists' insistence on linking Darwin to eugenics is quite clever, and probably effective. Though Darwin was not the originator of social Darwinism, creationists have effectively tied his name to the idea by citing eugenics as a logical extension of social Darwinism. Eugenics was an ugly pseudo-scientific concept that justified sterilizing people, based on vague criteria of fitness. In the early twentieth century, evolution and genetics were politicized via the eugenics movement, which wedded the science of genetics to the social and political issues of immigration and welfare. Eugenics was, in many respects, a logical extension of Herbert Spencer's application of natural selection to social evolution, which he called social Darwinism. Advances in the understanding of heredity in the late nineteenth and early twentieth centuries inspired the movement and made possible a scientific-sounding argument for the necessity of limiting the breeding of less desirable individuals—often identified by country of origin, ethnic group, or race. Leonard Darwin put the family imprimatur on eugenics when he dedicated his 1926 book, *The Need for Eugenic Reform*, to his father, Charles. The eugenics movement swept the country in the 1920s, with sterilization laws on the books in 24 states. In 1927, the U.S. Supreme Court, in *Buck v. Bell*, upheld a Virginia sterilization statute by an 8–1 vote, with Chief Justice Oliver Wendell Holmes delivering the memorable line: "Three generations of imbeciles are enough." Eugenicists attempted to distance themselves from the Nazi movement in the 1930s. The Nazis took eugenics to its logical extreme. Donations to the movement dried up with the onset of the Great Depression, and war meant the disappearance of donations.[63]

Beyond the walls of the museum, in an AiG publication, *Charles Darwin: His Life and Impact*, the ministry equated Darwinism with racism under the following subhead "The Seeds of Racism":

> This theory [evolution] logically implies that certain "races" are more ape-like than they might be human. Ever since the theory of evolution became popular and widespread, Darwinian scientists have been attempting to form continuums that represent the evolution of humanity, with some "races" being placed closer to the apes, while others are placed higher on the evolution scale. These continuums . . . are still used today to justify racism. . . .
>
> . . . The fruits of Darwinian evolution, from the Nazi conception of racial superiority to its utilization in developing their governmental policy, are well documented. . . .
>
> . . . Currently, members of the Ku Klux Klan justify their racism on the basis that they are a more evolutionary [*sic*] advanced race.[64]

AiG was correct about the use, or, more accurately, the misuse, of Darwinism to justify racism and eugenics in twentieth-century America and Europe.

The antievolution movement had on its side a sophisticated communications network, its own museum and plans for another, various institutes, and the allegiance of numerous national political figures. Its arguments, though, remained freedom of speech, populism, preservation of tradition, and a rebel cause, all proven tactics from Bryan's political playbook.

8

Legacy

Here has been fought out a little case of little consequence as a case, but the world is interested because it
raises an issue.

—William Jennings Bryan

I think this case will be remembered because it is the
first case of this sort since we stopped trying people in
America for witchcraft.

—Clarence Darrow

The Scopes trial was the first public performance of modern America's science-versus-religion drama.[1] Its high visibility and dramatic quality
gave it a special place in the subsequent fight because the trial defined terms
and tactics that have endured into the twenty-first century for the antievolution movement. Creationists still use Bryan's arguments against evolution
and his appeals to American myths and democratic values. His lessons in
practical politics and using the press to promote one's agenda have not been
lost on the modern antievolution movement. First, putting the fight in court
meant a public argument. Even without the impetus of sensation and the
bizarre, news media were obligated to cover courts, much as they did legislatures, school boards, and other public bodies. Those were familiar venues,
facilitating coverage and making the issue easier to follow. Second, defining
belief in evolution as a test of one's faith brought large numbers of religious
people into a fight they otherwise might have ignored. Eradicating nuances
helped enlist adherents, as it would in any campaign. People, by nature, will
shun the effort required to sort through, perhaps to unsatisfactory conclusions, the complexities of a topic such as the impact of modernism on theology, or the place of materialistic science in one's faith, or if science even has
a place in faith.

Conflict and spectacle inflated the trial's significance, just as they have fueled the contemporary campaign against evolution. Like the Scopes trial, the
Creation Museum and the Ark Encounter provide spectacle for creationists
and conflict for media, amplifying the controversy and narrowing the issue.
Such a battle has helped creationists by winning attention, making the issue comprehensible at a different level, and helping distill the antievolution
campaign into a simple dichotomy of God versus evolution.[2] The trial also
put creationists on the offensive, shifting their efforts from simply opposing
the teaching of evolution to eventually appealing to egalitarian values and
demanding equal time.[3] This imbued post-Scopes creationists with an aura of
tolerance, which they contrasted to the apparent exclusiveness of the scientific
establishment. The creationist appeal to tolerance fit not just a liberal western
tradition but a religious tradition as well. The principle of noninterference
in others' religious activity, according to historian Mark Silk, is "a well-nigh
unassailable point of view in American public discourse."[4] Since the Scopes
trial, creationists have invoked the principle to good advantage by insisting
that freedom of religion was at stake when they questioned the validity of
evolution. Mainstream media accommodated their challenge because tolerance melded neatly with press values of fairness and balance.

The Scopes trial focused the fundamentalist movement, which previously
was a denominationally disparate array of antievolutionists and antimodernists. The battle became literalism versus evolution, surrogates for the larger,
and less well-defined, concepts of traditionalism and modernism. The trial
defined the terms as science versus religion, thereby keeping it simple and
facilitating press coverage. Just as important, the trial was the stage for two
characters who would impress their interpretations and ideas on subsequent
generations of combatants. Though Bryan needed something to reverse his
flagging political fortunes, he was not merely opportunistic in looking at
evolution as an idea to give coherence to his objections to the modern world.
He sincerely did see in social-Darwinian–inspired militarism a threat to the
morality of civilization. In defense of Christianity and tradition, Bryan appealed to national myths of egalitarianism and individualism, and to Jeffersonian ideals of the popular will. Those values did not win a legal decision in
1925, but they fired creationist campaign rhetoric for the rest of the century.

Darrow, too, drew on national myths, especially the ideals of individual
rights and a rebel tradition that elevated individual dignity above a monarch's
comfort, materially or philosophically. The admiration of rebels is a national
myth that goes back to those who incited revolution against Great Britain.
The rebel impulse has run the gamut from great tragedies such as the Civil

War to moral triumph, as in the civil rights movement of the 1960s. No matter where one's ancestors may have been in an initial confrontation, the broader culture often embraced the rebels via such machinations as the "Lost Cause" apologia.[5] In all cases, though, individual rights eventually came to the fore in the rebels' discontent. Darrow personified that tradition.

Creationist Icon

The Scopes trial gave modern fundamentalists an iconic figure, even if they faulted Bryan's "liberal" theology, which meant his day-age accommodation. There were other figures for the office of chief fundamentalist in the 1920s, and some had better credentials in terms of being more hard-line literalists or absolutely antievolution, such as George McCready Price.[6] But the movement had something more in Bryan: a model for political engagement. Individuals such as Price and Bryan had many values and religious ideas in common. But Bryan offered a way to be mainstream, progressive, and rebellious by combining populist politics and anti-intellectualism with individual rights, all the while in opposition to an elitist, exclusive "club" of mainstream scientists. An experienced campaigner, he even provided an alternative definition of science. His basic argument was that "true" science was true to facts. The nature of truth has been the underlying argument in creationists' fight against science. For creationists, the Bible is the foremost book of facts, and thereby not subject to refutation. For Bryan, evolution could not possibly be true because it could not be congruent with the supernaturally guaranteed truth of the Bible. Truth was, from the creationist perspective, eternal and unchanging. Hence, "true science" could not be at odds with "Truth." That truth is fixed and eternal is necessarily irreconcilable with the idea that knowledge is fluid and changing.[7] Thus, Darwin and his mere "hypothesis" and "guesses" were not true science. Contemporary creationists, with the idea of intelligent design, have expanded their intellectual arsenal to include irreducible complexity and complexity by design.[8] Those armaments have fizzled scientifically, but they have sufficient sparkle for campaign purposes.

In the 1920s, evolution was doubly damned in Bryan's eyes. Not only did it fail as "true" science, many people disapproved of it. For Bryan, accepting evolution would have violated the sanctity of the Jeffersonian principle of the people's will. By insisting on the popular and Scriptural truth of their science, fundamentalists could see themselves as remaining progressive and not being relegated to the intellectual middle ages. They touted themselves as progressive by virtue of adhering to democratic principles. People had the right to decide their fates, whether manifest in the selection of officeholders or of scientific facts.

Like Bryan, modern fundamentalists understand how to win media attention, and have grasped very well the central role of media in creating and sustaining a constituency. The necessity of simplicity was another important lesson—draw more people into an issue by simplifying it and making it immediately relevant to them. The journalist Ray Stannard Baker condescendingly said of Bryan, in the 1912 Democratic convention in Baltimore, that he reflected an attribute of great leadership: the ability to make issues clear, "so common men who do not think may vote as they feel."[9] An enduring example of Bryan foregoing nuance is his treatment of the words "fact" and "theory." He erroneously presented them as antonyms. Theories organize and explain facts, but Bryan was adept at taking advantage of people's ignorance of scientific terminology and process. He deemed the word "hypothesis" a term that was "euphonious, dignified and high sounding," simply a "scientific synonym for the old-fashioned word 'guess.'" His rhetoric stuck and became a common fallacy among antievolutionists. Bryan also seized upon some scientists' reservations about any part of Darwin's theory to discredit evolution, as though a theory needed 100 percent acceptance and incontrovertibility to be valid and reliable. When the British biologist William Bateson, in his 1921 address to the American Association for the Advancement of Science, noted that natural selection had not been definitively demonstrated as the mechanism that drove evolution—not at all a controversial statement—it became in Bryan's words yet another piece of evidence that "every effort to discover the origin of species has failed."[10] In the twenty-first-century corollary, creationist ministries find people with advanced degrees to reject or criticize evolution. This, they charge, proves the theory's weakness and demonstrates its lack of acceptance among scientists.

Bryan's ideas about evolution were jumbled. He believed evolution valid for plants and animals but not humans. He accepted the idea of an old Earth, which was necessary for the process of speciation via natural selection. But he sided with biblical literalists espousing a young Earth, and he publicly condemned theistic evolution as a deadly compromise of religious principles.[11] "Darwinism" and "social Darwinism" were the same, in his estimation. *Theory, hypothesis, guess, speculation*—all were synonymous in his lexicon. Bryan mixed up the distinction between origin of life and speciation, the evolutionary formation of new species. Sweeping away such details cleared the path for his political campaign against evolution. He posed the issue most severely: "The language I have quoted proves that Darwin is directly antagonistic to Christianity. . . . Darwin, by putting man on a brute basis and ignoring spiritual values, attacks the very foundations of Christianity." When it came time for the national showdown, Bryan claimed his spot in the theological Armageddon: "They came to try revealed religion. I am here to defend it."[12] The tactic persisted because it worked—not scientifically, theologically, or philosophically, but politically.

Bryan's constituents were distant—often geographically, almost always intellectually—from the universities, those lairs of elitism. For Bryan, professors who embraced modernism and Darwinism were the personification of arrogant tyrants. He believed democracy and individual rights were at risk because science and religion would be subjugated to the presumed superiority of self-anointed intellectuals, willing to impose a materialist doctrine on school children across the nation. When Bryan embraced science and technology while dismissing evolution and modernism, he remained true to the American ideals, born in the nineteenth century, that made science and technology synonymous with progress. In "The Origin of Man" speech, Bryan pointed out that "medicine is one of the greatest of the sciences and its chief object is to save life and strengthen the weak." He set that against Darwinism, which he said aimed to eliminate the weak. Were Darwin alive, Bryan said, he surely would be appalled by the remedies for typhoid, yellow fever, and the black plague, as well as efforts at vaccination and attempts to cure tuberculosis and cancer. "Can such a barbarous doctrine be sound?" Bryan asked.[13] And so he was able to damn Darwin and praise science.

Bryan was pragmatic not just with science, but also with religion. In a very American tradition of pragmatism, Bryan looked to Christianity for practical consequences for individuals and societies. Theological fine points were not his primary interest. Christianity, he believed, benefited culture and should be used to better society. In this way, his progressive politics and Christian principles melded neatly.[14] This paralleled his ideas about science, in which the more immediate concern was the impact on people. In this respect, Bryan was very much in concert with the larger culture, where the love of science was not with theory but with the practical results—a perspective that went back to Benjamin Franklin.[15] Christianity had practical results: it made people and society better. Darwinism degraded both, in Bryan's opinion. Like religion, science was important for its application, and just as one would discard religion that harmed people, so would one discard science that had bad results for society.

Bryan made the evolution threat concrete and immediate. He fit evolution to a specific grievance, which was the abandonment of traditions both sacred and secular. He made it urgent, in need of redress, when he designated public schools as the battleground. Bryan successfully turned the fight against modernism, an intellectual movement, into a populist campaign issue. Likewise, twenty-first-century fundamentalists elevated Darwin to a special place among offenders against the truth and tradition, bundling evolution with other threats such as abortion, homosexual marriage, and the lack of prayer in schools.

Inherit the Wind's caricature of Bryan has backfired. The movie's Bryan was modeled on Mencken's parody—a vain, bloviating fool with a large following of like-minded fools. The depiction metastasized in public schools as drama teachers picked up the play, and textbooks exploited the liberal disdain for evangelicals in the person of Bryan.[16] But Bryan's successors nurtured the resentment engendered by the film and have won nearly half the American public in the decades-long antievolution campaign. The contempt of Mencken and other intellectuals may have helped Bryan's constituents identify themselves by defining more clearly what they were not—urbanites, modernists, elitists. The simplistic history in *Inherit the Wind* benefited fundamentalists by diminishing history. The play and movie were not made as historical documentaries, but the historical context for the indictment of intolerance is inescapable. Small history benefits small thinkers. A one-dimensional view of the Scopes trial is to the advantage of those whose cause would be hindered by complexity and nuance. Just as *Inherit the Wind* created fundamentalist backlash, the Scopes trial embittered many fundamentalists because they felt they had been derided and marginalized. But the derision was counterproductive. Before the trial, antievolution was not a major issue for fundamentalists, but it became one afterward as literalism became their central tenet. Events in the 1920s were at the beginning of a trend in which fundamentalism existed in a symbiotic relationship with aggressive liberalism or secularism. When attacked, fundamentalism became more extreme and bitter.[17] Just as conflict has shaped the identity of fundamentalism, so has it aided the intelligent-design movement. In democratic politics, conflict polishes and refines the issues. For creationists, conflict enhanced a story line of heroes and villains for consumption by media. Confrontation, however, has not been mere manipulation of public opinion and media. After all, how could one compromise on such a thing as the nature of ultimate truth? When fundamentalists reduced the conflict to familiar symbols of good and evil, they made family the focal issue of their political fight. Just as "modernism" would have been a tougher enemy to identify for most people in the 1920s, so would "secular human-ism" be a problematic sale in more recent times. However, evolution's threat to the traditional, biblical family roles is discernible and subsumes a number of other issues.[18] Politically, positioning oneself with respect to a single issue is much simpler than dealing with myriad issues, particularly when the issue is cast so as to make disagreement politically untenable. In other words, what candidate would dare to be "antifamily"?

In an age of mass media, celebrity status was critical to winning attention for a cause. Bryan had recognition—then and now. He entered the trial as a celebrity politician and fundamentalist leader, and history has granted him the role for posterity, thanks in large part to *Inherit the Wind*. Creationists have attempted to re-create a Bryanlike icon and celebrity. Such an attempt was evident in *Darwin's Nemesis*, which tried to make an icon of Phillip Johnson, who was credited with being the intellectual fire behind intelligent design. The essays in the book laud Johnson not only for his Christianity, but for being an antiestablishment rebel and an individualist.[19] The book even devoted a chapter to Johnson's antithesis, Richard Dawkins, who apparently was designated to be a sort of Darrow to Johnson's Bryan.

Darwin's Nemesis went to creationists' core issue concerning the purpose of education—the Bible as the foundation of all knowledge versus an empirical grounding that is divorced from religion.[20] That segment of the book pitted Dawkins and Dennett against Johnson and was reminiscent of the Bryan-Darrow courtroom debate, only this time framed from the Bryan-Johnson side. *Darwin's Nemesis*, along with the *Wedge Document*, revealed a movement that is essentially political: a campaign to change public policy is the reason for its existence; the movement has a figurehead; there is a strategic plan and call to action. The movement's defining events—publication of *Darwin on Trial*, and the 1996 Biola University conference that spawned the wedge strategy—are campaign/policy driven. One identified a "founding father," and the other a plan for enacting his vision of establishing a Christian culture.

Evolution's Icon

In the course of defending Scopes—or, perhaps more accurately, prosecut-ing religious tyranny—and especially in his summation, Darrow used argu-ments that have been resurrected with each new episode of the debate. One of the most enduring themes has been the irreconcilability of science and religion. This dichotomous approach was common in the 1920s debate. This "either-or" tactic was familiar to Darrow, especially in his labor cases, where it was either labor or capital. He was uncompromising in this respect and in his lawyerly use of war images, in which one side must prevail and the other fall. More significant, though, in the continuing controversy was his emphasis on individual rights, which is still a common theme among anti-evolutionists. In the context of the Scopes trial, Darrow meant not just the right to be free to teach evolution, but the right to be free of religion and not compelled to support a set of beliefs. Darrow's contemptuously dismissive stance toward literalists is the same one adopted by modern scientists toward creationists such as Johnson and Behe, whose antievolution books are aimed at audiences unversed in science and unconstrained by logic. Scientifically, Darrow's disdain was merited, but he missed the point: Science was less the issue than winning public attention and status with a well-defined constituency—which did not include mainstream scientists and their allies. In the twenty-first century, many scientists still have not grasped this important distinction, while creationists have shucked the lab coat to ascend the stump.

Darrow insisted on empirical evidence, the imperfection of creation, and humanity's susceptibility to delusions: "Man is not only the home of microbes, but of all sorts of vain and weird delusions. . . . Man really assumes that the entire universe was made for him; that while it is run by God it is still run for man."[21] Darrow's agnosticism and his disdain for religion may have failed his own cause because he underestimated the adversary. A Menckenesque dismissal of the opposition with a bit of name calling may have entertained a sympathetic audience, but it only infuriated a hostile one, while only briefly winning the attention of a largely indifferent populace. But such an attack did not engage the ideas and beliefs at the core of the opposition. Instead, Darrow tended to respond to the antievolutionists, rather than initiate the campaign. He was both reactionary and dismissive. Those aspects of his legacy are intact in the scientific side of the contemporary debate.

In contrast, Bryan did not debate; he campaigned. Bryan did not appeal to intellect, but to values and traditions. In the contemporary fight between creationism and science, creationists have adopted Bryan's politics and part of Darrow's appeal. It is a populist campaign, in the Bryan tradition, but fundamentalists have taken a cue from Darrow, casting themselves as the folk-rebels, the aggrieved minority in which individual rights are preeminent. The philosophical differences between Darrow and Bryan are the same differences that separate creationists and scientists in the twenty-first century. The Scopes trial remained a reference point for the evolution-religion fight because it was the clearest demonstration of this Kantian-Baconian conflict. Now, the approaches can be seen most starkly by materialists such as Dawkins and Dennett, and fundamentalist-creationists such as Ken Ham and his Answers in Genesis ministry. Bryan's intellectual descendents have adhered faithfully to his Baconian concept of a rational world organized by God. For them, the job of science was and is simply to organize knowledge in the context of that religious frame.[22]

Both Bryan and Darrow taught subsequent generations something about conducting public battles over complex issues. Bryan showed constituents how to simplify the issue so that it was not modernism, but evolution. Then simplify the issue further, speaking for the Bible without discussions about day-age theory versus 6,000-year-old Earth, and simply equate evolution with social Darwinism and atheism. Also, speak to the masses and forget the elite, or use them as foil. The approach resonated in a culture with an anti-intellectual tradition. All the while, embrace values common to America such as Jeffersonian democracy, individualism, agrarianism, and admiration of science and technology. In similar fashion, Darrow simplified the issue, narrowing religion to biblical literalism. He appealed to the audience by speaking to the middle and marginalizing the opposition. But, most of all, he appealed to individual rights.

Darrow established a logical approach for contemporary materialists in *Story of My Life*, where he argued that God is not an explanation of the material world and that some things are just not scientific questions. The conception of immortality is simply a 2,000-year-old piece of ignorance handed down from primitive nomads, he stated. The parallels to contemporary rationalists are not just in the substance of the argument, but in a tone that anticipates the likes of Dawkins and his acidic eloquence. Darrow observed:

> The conception of immortality, so far as the Western world is concerned, came from the ignorance of two thousand years ago. A primitive tribe of nomads believed that the whole universe was made for them. . . . The universe had been specially created for them.[23]

Dawkins, more recently, voiced a similar criticism:

> We're talking about a tribe of wandering Middle Eastern herdsmen. Why would they have any wisdom about the origin of the world or the origin of anything else? That particular myth, the Judeo-Christian myth, happens to have come to our civilization.[24]

In similar Darrowinian fashion, Dawkins charged, "Among the dishonesties of the well-financed intelligent-design cabal is the pretense that the designer is not the God of Abraham but an intelligence unspecified, who could equally well be an extraterrestrial alien."[25] Like Darrow, Dawkins was sharply analytic, cutting through the opposition's fog to the central issue, unleashing a logical fury on a "dishonest cabal," and casting religion into the heap of historical superstitions. Dawkins was one of several public intellectuals who has carried the Darrow tradition into the twenty-first-century battle against state religion and ignorance in the name of God. Other public intellectuals in that vein have been Christopher Hitchens, who died in 2011; Michael Shermer; and Daniel Dennett. But none have matched Darrow's unique ability to engage his religious adversaries and simultaneously hold the moral ground in defense of individual rights and learning, all the while antagonizing and charming with wit.

The Trial Myth

The durability of young-Earth creationism should dispel the idea that the Scopes trial effectively defeated creationism when Darrow bested Bryan. *Inherit the Wind* gave that legend new energy with Brady/Bryan collapsing theatrically and incoherently at the end of the film. For the most part, fundamentalists ignored their apparent demise. Insulated in their own religious communities and academies, fundamentalist-creationists did not have to confront challenges to their faith and literalism. Fundamentalists faded from the front pages after Scopes, but they remained active, especially in rural and small-town America, fighting evolution in school textbooks, classrooms, and curricula.[26] The harsh denunciations by Darrow, Mencken, and others only aggravated fundamentalist fear and alienation. Even before Scopes, fundamentalists saw a godless world looming, and it frightened them. Reactionary movements, which have existed not just among Christian fundamentalists but also Moslem and Jewish fundamentalists, are grounded in fear and anger. The Scopes trial left fundamentalists even more convinced the goal of secularists was to eradicate them.[27] The trial was a unifying event because it narrowed and focused fundamentalists' grievances.

Antievolutionists remained vital and campaigned energetically and successfully for textbook changes and constraints, minimizing evolution and avoiding controversy.[28] The fundamentalist defeat is the basic falsehood and legend of the trial. The loss myth focused not on the verdict but Darrow's humiliation of Bryan—and, therefore, fundamentalism—on the stand, and Bryan's subsequent death. But, over time, Bryan's apparent loss focused the fight against modernism and provided an icon for creationists, even if they disagreed with some of the particulars of Bryan's antievolutionism. In one respect, the fundamentalist triumph was demonstrable. Mention of evolution in school textbooks dropped dramatically after the trial, and it was the 1960s before evolution began to return to textbooks. The rush to diminish evolution in the textbooks was so dramatic after the trial that, by 1930, 70 percent of high-school texts dropped the subject altogether.[29]

The Scopes trial was inconsequential as legal precedent, but monumental as a defining point for the evolution-creation conflict. A conservative religious movement found in evolution an adversary that people recognized, could relate to personally, and that encapsulated their fears of the modern world. Not only had fundamentalists found a foe that was unpopular to begin with, but religion grew in the decades after the trial. Up to and after World War II, and with the threat of "godless communism" in the 1950s, Darrow's view of religion as fraud fell even further into disfavor. Church attendance climbed while secularism declined in public respectability for several reasons. First, mainline churches simply accommodated secularism by reconciling science and religion via an allegorical interpretation of scripture. People did not have to deny God in order to accept science. In addition, the taint of elitism politically weakened strong secularists in American universities.[30] Politically, Bryan's formulation had shown itself to be a good one: appeal to the masses, denounce a lack of faith, and marginalize the opposition as elitists operating in an antidemocratic fashion.

Useful Myths

The "truth" about America, according to historian James Oliver Robinson, resides both in its mythology and its history.[31] A myth may or may not be historically true, but as a story it teaches a lesson, transcending events and details in order to understand a greater truth. Myths clarify issues and explain problems, often an issue that cannot be solved by reason or logic. A myth resolves the tension and solves apparent contradictions in reality—i.e., how ancient fossils can exist in a 6,000-year-old Earth—for those who embrace the myth.[32] Myths not only address anomalies in world views, but they simplify complex issues, such as American idealization of and commitment to individualism, or the conquest of new frontiers, whether geographic or scientific.[33] Four that are consistent in literature on American myth and are particularly appropriate to creationism/intelligent design are the garden, the frontier, progress and science, and individualism and egalitarianism.

When Darwinism and modern science confronted the idea of Eden's literal existence, agrarian America, particularly the South, set itself against the urban-industrial Northeast. With the help of journalists such as Mencken, the ignorant, country fundamentalist became a national stereotype. The journalistic representatives of urban American were not solely responsible for characterizing fundamentalism as a rural or small-town phenomenon. Bryan, too, framed the issue as one of Jeffersonian agrarianism and fundamentalist Christianity against the corrupting forces of modernism, which lurked in cities and universities. The suspicion of universities and belief in the garden continues among contemporary creationists. Evolution still remains a threat—the designated serpent—in the creationists' garden.

The antievolution movement has shrouded itself in apparent open-mindedness and adventure, claiming to venture intellectually to places—such as a 6,000-year-old Earth and 7-day creation—from which they accuse mainstream scientists of shrinking. Their self-proclaimed adventurousness slips easily into the national frontier myth. In earlier times the reference point was the vast North American wilderness, to be conquered, and out of which utopia would be carved. The frontier myth changed over time to represent opportunity, whether for wealth, democracy, or paradise. Like the West of earlier years, the city also could be a frontier, a wilderness of savages in the form of immigrants and other strange people, but a place in which to expand democracy and faith. Mission and wilderness were directly linked in American history.[34] The frontiers to be conquered have ranged from geographic ones, including the West and outer space, to social ills (such as Lyndon Johnson's "Great Society"), to science and medicine (witness our multitude of "wars" on specific diseases, such as cancer and addiction). John F. Kennedy even invoked the symbol in his "New Frontier" speech in his 1960 inaugural address. Intelligent-design proponents have seized on this mythic theme by casting themselves as the bold ones, the ones seeking to discover "new worlds," places that timid conventionalists—i.e., mainstream scientists—dare not go for fear of offending colleagues also mired in the machinery of convention. In such a scenario, the shunned outsiders turned into frontiersmen, trekking into intellectual frontiers.

The frontier myth campaign meant creationists could move in concert with, rather than in defiance of, another American myth: science as progress. This mythology accelerated in the nineteenth century as belief in science merged with belief in progress. Darwinism did present a problem for American Christianity, but it was partially reconciled by separating science and religion in the nineteenth century. As a result, an appeal to science and technology in American society commonly was associated with an appeal to progress and forward-looking ideas. It is part of an older myth of rationalism, what Robinson called "finding law," or the idea that true laws exist and, with evidence and logic, can be discovered: "And as much as medieval peasants and knights believed themselves utterly subject to the laws of God, so Americans believe themselves subject to the laws of science."[35] This created tension with fundamentalist Christianity. In the case of intelligent design, recognition of the status of science is seen in allusions to data and theories, and the intelligent-design use of scientific language to demonstrate the progressivism of their brand of creationism. It is a lesson learned from the book of Bryan, who showed creationists how one could be true to science and progress while denouncing Darwin: Truth could not be at odds with itself because "true science" could not be at odds with Scriptural truth. With this same logic, contemporary creationists claim to be on the side of true science versus mere materialism, which excludes an absolute or higher truth.

Frontiers and progress can be substantial myths only if there are individuals with the force of character to place the values in the national experience. Egalitarianism is inherent in the individualism that permeates American culture. Individualism involves strength of character, defying authority or establishment, being a pathmaker or a trailblazer. One can't be static and part of a bureaucratic structure in this particular myth. According to Robinson, "A free American in pursuit of happiness . . . is *mobile*, is, has been, will be, in motion *and* is defying authority, pathmaking (trailblazing), is in fact revolutionary. Moving, trailblazing revolution in defiance of establishment is the *way* of the proper American individual" [emphasis in original].[36] An important variation on this myth is the love of the underdog, or Horatio Alger in its American form. It is the individual who overcomes the odds. Being enamored of the underdog is as ancient as David and Goliath, and it is evident in American myth in characters as diverse as Benjamin Franklin, Abraham Lincoln, and Bill Clinton. Their mythologies all include humble origins. A consistent theme in news stories—and one exploited by the intelligent-design proponents themselves—is the creationist who takes on the privileged, establishment scientists.

Egalitarianism and individualism, according to Robinson, have been elevated as the logic of federalism declined in American politics, to be replaced by a progressive ideal of equality: "[E]ach individual has a vote which entitles him or her to representation in government, rights before the law, and a proportion of the community's or the nation's advantages equal to those of every other individual."[37] In 2005, the Kansas state board of education exercised a similar rejection of the "federalism" of mainstream scientists by altering the definition of science in order to accommodate intelligent design. Bringing the debate to a political arena creates an inevitable tension between science and democracy that intelligent-design proponents have exploited very well.

Ideally, democracy is egalitarian with regard to individuals; science does not treat all ideas equally. Creationists have criticized scientists for being authoritarian or undemocratic for not entertaining intelligent design. The critique wrongly presumed that ideas, like people, are created equal.

Using these mythic themes, antievolutionists are writing their own version of history. They have redefined science in the spirit of Bryan to make it a political, rather than intellectual, endeavor. This rewriting of history and science thrives, and will continue to do so, because literalists have successfully created what two scholars call a parallel culture in America. It is a culture with its own education system, from home schooling to colleges and universities; its own publishers; its own, near-closed social circles; and its own media system that includes print, web, and broadcast outlets for filtering information from the world outside. Within this parallel culture, it becomes easy to believe in the absurd—i.e., Earth is 6,000 years old, people and dinosaurs coexisted, a flood explains geologic strata, and so on—because its own history, religion, and science simply preclude the alternatives.[38] The creation of their own history and science provides literalists a way of linking diverse fundamentalist denominations to one another and to the larger American culture and tradition. The allure of a romanticized past is a powerful draw in unsettled times, especially for a group that has seen itself under threat by forces such as secular humanism. Linking to American history legitimizes the movement, and focusing on evolution identifies its demon. These twin prongs of the creationist movement's identity are much in the Bryan mold. Scopes and Bryan are central to contemporary creationism because the event and the man provided antievolutionists a historical narrative of injustice and righteousness, grounded in American cultural mythologies and values. For the literalists, the trial is a historical reference point, legitimizing creationism as a force in American history.

Science versus Symbols

Growing out of an evangelical tradition, creationists have a familiarity with and skill in appealing to symbols, myths, and a powerful poetic tradition. They are part of a tradition that quickly embraces new media forms—radio, television, the internet—and molds the message to the strength of the media.[39] As the latest new media technologies emerged, evangelists have adapted them to existing organizations and strategies. The ministry already had a central purpose, a basic message, and skilled communicators. Their mission, after all, was to go into the world, however and wherever one found it, and spread the word, whether door-to-door or network-server to network-server. Using symbols and myths is a daily practice for evangelists. Such is not the case for most professional scientists. Often, the problem for scientists is to maintain the role of expert, which the public respects, and which implies being apart from common knowledge or discourse. Such implicit elitism may be unappealing to the general public. Few scientists are able to be both defender of common good and a public intellectual. Darrow could do it. Bryan could not, as he was defender of common man, but assailed the elitist intellectuals. Therein lies another problem for scientists in a modern, democratic society—to be expert and elite while addressing the most common parts of an egalitarian culture.

For the most part, scientists are unaware of their decline in the campaign for public opinion, primarily because they do not understand the strength of the opposition, which is essential to taking effective action.[40] In the contemporary debate, creationists understand and use media. Scientists understand science, and argue the issue from perspectives of validity and reliability, which the public largely disregards or does not understand. Scientists tend to be apolitical about their profession, which is in contrast to the moral citizenship endorsed by Bryan and embraced by creationists and conservative Christians. Scientists are at a disadvantage in this contest because their primary goal is to do science. The primary goal of creationists is to evangelize, and science is of concern only inasmuch as it is a tool in promoting the primary message, which is religious. Scientists have continued to make a logical, empirical, scientific case in order to vanquish those who have no regard for logic, empiricism, or science.[41]

The Creationist Paradox

The media face a dilemma in covering any number of issues that appear to be anything from merely odd to blatantly outlandish: the issue is newsworthy, even if not morally, politically, or scientifically legitimate. Contemporary creationists' campaigns against evolution are scientifically bankrupt but politically ingenious because making the issue a political problem meant a political solution must exist. If it is a political problem, then the standards and expectations of political news apply: impartiality and balance.

Creationists and intelligent-design advocates have succeeded in being labeled "conservative," a political designation putting them in the national political discourse, part of a grand debate that has endured for the life of the republic. They are not fringe or oddball cultists when they are part of the two-party system and a debate over social values. Lawsuits, school-board debates and legislative action create news. The issue becomes a public issue, no matter the lack of scientific merits. The success of creationism is apparent in the fact that a national debate even exists over an issue about which there is no scientific controversy. It also is a testament to political skill. Phillip Johnson declared, "It is nearly inevitable that 'teach the controversy' will become public policy."[42] As if to prove him right, the ardently empirical and proevolution *Skeptic* magazine ran an article in 2009 titled, "It's time to teach the controversy: Since creationism isn't going away, let's use it in the classroom to teach the difference between science and pseudoscience."[43] It was not what Johnson had in mind, but even the appearance of such a suggestion in a rationalist publication showed the triumph of tenacity, if nothing else.

Creationism and intelligent design will not vanish anytime soon. The movement is tied intimately to American history and myth, and it is tactically modern and innovative with the campaign to bring God back to the secular realm.[44] Creationists have manipulated their history, transforming humiliation in a backwoods Tennessee village into a national movement. The manipulation, though, has not been capricious. Creationists have factually filtered their history and made it congruent with deeper cultural values in order to fit a narrative of righteous rebels fighting for individual liberty and freedom of expression. Creationists have the ministries, academies, and institutes to generate the material, the communications network to disseminate it, and the constituency to support it. Creationists will do with history what they have done with science: proclaim the one great "truth," and see that all the facts fit it. In such a narrative, it is ironic that creationists are fighting to win intellectual respectability by assuming the trappings of their opposition—scientists. It is doubly ironic that they are succeeding.

Notes

Introduction. Creationism's Political Genesis

1. Ronald L. Numbers, *The Creationists: From Scientific Creationism to Intelligent Design* (Cambridge, Mass.: Harvard University Press, 2006), 369–370.

2. Chris Mooney, *The Republican War on Science* (New York: Basic Books, 2006).

3. Barbara Forrest and Paul R. Gross, *Creationism's Trojan Horse: The Wedge of Intelligent Design* (New York: Oxford University Press, 2004).

4. Eugenie C. Scott, *Evolution vs. Creationism: An Introduction* (Los Angeles: University of California Press, 2004).

5. Michael Ruse, *The Evolution-Creation Struggle* (Cambridge, Mass.: Harvard University Press, 2005), 261, 287.

6. Edward Larson, *Summer for the Gods: The Scopes Trial and America's Continuing Debate over Science and Religion* (New York: Basic Books, 1997); Ray Ginger, *Six Days or Forever? Tennessee v. John T. Scopes* (Boston: Beacon Press, 1958), 41, 222.

7. Michael Lienesch, *In the Beginning: Fundamentalism, the Scopes Trial, and the Making of the Antievolution Movement* (Chapel Hill: University of North Carolina Press, 2007).

8. George M. Marsden, *Fundamentalism and American Culture: The Shaping of Twentieth-Century Evangelicism: 1870–1920* (New York: Oxford University Press, 1980).

9. Matthew Chapman, *40 Days and 40 Nights* (New York: Harper Collins, 2007); Laura Lebo, *The Devil in Dover: An Insider's Story of Dogma v. Darwin in Small-Town America* (New York: The New Press, 2008).

10. Edward Humes, *Monkey Girl: Evolution, Education, Religion, and the Battle for America's Soul* (New York: Harper Perennial, 2007), 258.

11. Marcel C. LaFollette, *Creationism, Science, and the Law: The Arkansas Case* (Cambridge, Mass.: MIT Press, 1983); Edward Larson, *Trial and Error: The American*

Controversy over Creation and Evolution (New York: Oxford University Press, 1985); Dorothy Nelkin, *The Creation Controversy: Science or Scripture in the Schools* (New York: W. W. Norton, 1982).

12. Frank Ravitch, *Marketing Intelligent Design: Law and the Creationist Agenda* (New York: Cambridge University Press, 2011).

13. Tona Hangen, *Redeeming the Dial: Radio, Religion, and Popular Culture in America* (Chapel Hill: University of North Carolina Press, 2002).

14. Mary Beth Swetnam Mathews, *Rethinking Zion: How the Print Media Placed Fundamentalism in the South* (Knoxville: University of Tennessee Press, 2006).

15. On the Christian tradition of republicanism, see David Sehat, *The Myth of American Religious Freedom* (New York: Oxford University Press, 2011), 6–9, 21–22.

16. Forrest and Gross, *Creationism's Trojan Horse*, 273, 332–333.

17. On the differences between intelligent design and creationism, see Scott, *Evolution vs. Creationism*, 51–68.

18. On politicizing science in recent decades, especially as a phenomenon in the Republican Party since the 1980s, see Mooney, *Republican War on Science*.

19. www.gallup.com/poll/21814/evolution-creationism-intelligent-design.aspx, accessed October 10, 2011.

20. James Oliver Robinson, *American Myth, American Reality* (New York: Hill and Wang, 1980), 285–286.

21. Ibid.

22. Ibid., 281–282, 287.

23. Ibid., 128–129.

24. On the construction of parallel cultures, see Randall J. Stephens and Karl W. Giberson, *The Anointed: Evangelical Truth in a Secular Age* (Cambridge, Mass.: The Belknap Press, 2011).

25. Scientists on occasion engage the political aspects of these issues. See, for example, *New York Times*, "Groups Call for Scientists to Engage the Body Politic," August 9, 2011, D1; "Seeking a Missing Link, and a Mass Audience," May 19, 2009, A1.

26. www.gallup.com/poll/21814/evolution-creationism-intelligent-design.aspx, accessed September 25, 2012.

Chapter 1. The Genesis of Young-Earth Creationism

1. The quotes from Darrow and Bryan are from *The World's Most Famous Court Trial: State of Tennessee v. John Thomas Scopes* (New York: Da Capo Press, 1971[reprint of 1925 trial transcript]), 79, 338.

2. On the history of creationism and evolution, see: Michael Ruse, *The Evolution-Creation Struggle* (Cambridge, Mass.: Harvard University Press, 2005); and Ronald Numbers, *The Creationists: From Scientific Creationism to Intelligent Design*, expanded edition (Cambridge, Mass.: Harvard University Press, 2006).

3. Karen Armstrong, *The Battle for God* (New York: Alfred A. Knopf, 2000), 169–170.

4. David Sehat, *The Myth of American Religious Freedom* (New York: Oxford University Press, 2011), 185–187, 200.

5. Numbers, *Creationists*, 53; Michael Lienesch, *In the Beginning: Fundamentalism, The Scopes Trial, and the Making of the Antievolution Movement* (Chapel Hill: University of North Carolina Press, 2007), 11, 29; Armstrong, *Battle for God*, 170–171.

6. Ruse, *Evolution-Creation Struggle*, 161–162; Dorothy Nelkin, *Selling Science: How the Press Covers Science and Technology* (New York: W. H. Freeman and Company, 1997), 30; Edward Larson, *Summer for the Gods: The Scopes Trial and America's Continuing Debate over Science and Religion* (New York: Basic Books, 1997), 32. On the emergence of the WCFA, see Armstrong, *Battle for God*, 173.

7. Lienesch, *In the Beginning*, 19.

8. Numbers, *Creationists*, 53–55; Edward Larson, "Before the Crusade: Evolution in American Secondary Education before 1920," *Journal of the History of Biology* 20 (1987), 113.

9. Lienesch, *In the Beginning*, 83–87, 94.

10. Numbers, *Creationists*, 53, 57.

11. Clarence Darrow, *The Story of My Life* (New York: Charles Scribner's Sons, 1932), 88–92, 94–95.

12. Clarence Darrow, *Clarence Darrow on Religion*, "The Pessimistic versus the Optimistic View of Life" (Forgotten Books, republished 2007, www.forgottenbooks.org, accessed February 4, 2013, originally published 1910), 24.

13. Ibid., 165–171.

14. Quoted in Arthur and Lila Weinberg, *Clarence Darrow: A Sentimental Rebel* (New York: G. P. Putnam's Sons, 1980), 226.

15. On the relationship between Steffens and Darrow, see Weinberg, *Clarence Darrow*, 169–71, 225–226.

16. See Weinberg, *Clarence Darrow*, for a complete list of publications and speeches, 432–437. A variety, but by no means complete listing of publications is available at the web site for the University of Minnesota Law School, "The Clarence Darrow Digital Collection," http://darrow.law.umn.edu/trials.php?tid=19, accessed December 16, 2011.

17. Ibid., 111–112. The Rowell's numbers are on 331, footnote 13.

18. Lienesch, *In the Beginning*, 62.

19. Michael Kazin, *A Godly Hero: The Life of William Jennings Bryan* (New York: Alfred A. Knopf, 2006), 41, 185.

20. Ibid., 122, 131–132.

21. Lienesch, *In the Beginning*, 62.

22. Kazin, *Godly Hero*, 274.

23. Cited in Mary Beth Swetnam Mathews, *Rethinking Zion: How the Print Media Placed Fundamentalism in the South* (Knoxville: University of Tennessee Press, 2006), 68.

24. *Time*, May 12, 1923, "Who can be saved," cited in Mathews, *Rethinking Zion*, 70.

25. *Nation*, "The Orthodoxy of Democracy," June 6, 1923, 645; "The Case of Fundamentalists," December 26, 1923, 729; *New Republic*, "The Parson's Battle," January 9, 1924, 162.

26. Charles William Eliot, "The Great Religious Revival," *Atlantic Monthly*, March 1924, 379–385, cited in Mathews, *Rethinking Zion*, 71.

27. George M. Marsden, *Fundamentalism and American Culture: The Shaping of Twentieth-Century Evangelicism: 1870–1925* (New York: Oxford University Press, 1980), 5, 32–35, 130–135.

28. *Century*, "An American Looks at His World: Fashions in Bigotry," May 1923, 158. The Unitarian minister's remark is in Mathews, *Rethinking Zion*, 72.

29. Heywood Broun, "A Bolt from the Blue," *Nation*, July 21, 1920, 128.

30. John Dewey, "Fundamentals," *New Republic*, February 6, 1924.

31. Paul Murphy, *The Rebuke of Southern History: The Southern Agrarians and American Conservative Thought* (Chapel Hill: University of North Carolina Press, 2001), 121–123.

32. Mathews, *Rethinking Zion*, 81–83.

33. Cited in Mathews, *Rethinking Zion*, 85–86.

34. That Darwin is attached to the term *social Darwinism* is a historical misnomer. The real father of the idea of social Darwinism was the English philosopher Herbert Spencer, who extended Darwin's biological thesis to human society. This distinction often was lost in popular culture in favor of the more widely known Darwin. That remains the case. The classic historical work on the subject is Richard Hofstadter, *Social Darwinism in American Thought* (Boston: The Beacon Press, 1955). A later and closer analysis is Richard Bannister, *Social Darwinism: Science and Myth in Anglo-American Social Thought* (Philadelphia: Temple University Press, 1979).

35. *World's Most Famous Court Trial*, 299.

36. William Jennings Bryan, *Fighting to the Death*, "William Jennings Bryan's Last Message" (Dayton, Tenn.: William Jennings Bryan University, 1949), 5.

37. *World's Most Famous Court Trial*, 321–335.

38. Darrow, *Story of My Life*, 246, 249, 257.

39. T. V. Smith, *Scientific Monthly*, "Bases of Bryanism," May 1923, 505–507.

Chapter 2. The Contrarian and the Commoner

1. William Jennings Bryan, *In His Image*, "The Origin of Man," reprinted edition (Champaign, Ill.: Book Jungle, undated), 81.

2. *Chicago Daily Tribune*, "Darrow Doesn't Want to Die for Past or Future," December 24, 1917, 4.

3. Michael Kazin, *A Godly Hero: The Life of William Jennings Bryan* (New York: Alfred A. Knopf, 2006), xv, 188.

4. *Boston Daily Globe*, "Attack of Apoplexy Kills Commoner as He Sleeps," July 27, 1925, 7.

5. *Chicago Daily Tribune*, "Death Takes Bryan," July 27, 1925, 1.

6. *Washington Post*, "William Jennings Bryan," July 27, 1925, 4.

7. Arthur and Lila Weinberg, *Clarence Darrow: A Sentimental Rebel* (New York: G. P. Putnam's Sons, 1980), 91–95, 127–152.

8. John A. Farrell, *Clarence Darrow* (New York: Doubleday, 2011), 9, 38; Clarence Darrow, *Clarence Darrow on Religion*, "Facing Life Fearlessly," "Absurdities of the Bible," "Why I Am an Agnostic," "The Pessimistic versus the Optimistic View of Life" (Forgotten Books, republished 2007, www.forgottenbooks.com, accessed February 4, 2013, originally published 1910), 12–13.

9. Karen Armstrong, *The Battle for God* (New York: Alfred A. Knopf, 2000), 182.

10. Kazin, *Godly Hero*, xv, 43. Bryan probably spoke with some irony because voters, in a nonbinding preference vote, had selected him over his Republican opponent by a 73-point margin. At that time, though, state legislators selected U.S. senators, and Nebraska legislators did not prefer Bryan.

11. William Jennings Bryan and Mary Baird Bryan, *The Memoirs of William Jennings Bryan* (Chicago: John C. Winston Company, 1925), 9.

12. He probably meant the American Association for the Advancement of Society. The *Memoirs* were finished by his wife, and she may have been unfamiliar with the organization name.

13. Bryan, *Memoirs*, 526–528.

14. On Bryan's constituency, see Kazin, *Godly Hero*, 197–198.

15. Ibid., 59–61, 71, 82, 103.

16. Ibid., 34.

17. Ibid., 379, 123.

18. Andrew E. Kersten, *Clarence Darrow: American Iconoclast* (New York: Hill and Wang, 2011), 166–168.

19. Farrell, *Clarence Darrow*, 290.

20. Weinberg, *Clarence Darrow*, 10.

21. See ibid., 157, for an impressive list of the intelligentsia at one of the dinner parties.

22. Ibid., 351–352. On the club's membership, see Darrow letter to Mary Field, December 6, 1915; cited in footnote 28, 406.

23. David Sehat, *The Myth of American Religious Freedom* (New York: Oxford University Press, 2011), 55–57.

24. Bryan, *Memoirs*, 13. Bryan also grossly oversimplified the social Darwinism philosophy, which was not Charles Darwin's idea, but that of English philosopher Herbert Spencer.

25. A copy of the resolution is in Bryan, *Memoirs*, 459–460.

26. Michael Lienesch, *In the Beginning: Fundamentalism, the Scopes Trial, and the Making of the Antievolution Movement* (Chapel Hill: University of North Carolina Press, 2007), 59–60.

27. Ibid., 95–96, 109–113, 128–130.

28. William Jennings Bryan, *Seven Questions in Dispute* (New York: Fleming H. Revell Company, 1924), 10.

29. Royce Jordan, "Tennessee Goes Fundamentalist," *New Republic*, April 25, 1925, 258–260.

30. Bryan, *Memoirs*, 479.

31. Ibid., 479–486.

32. *Chicago Daily Tribune*, "Bryan Runs Again; Seeks Church Post," May 7, 1923, 7; "Bryan Refuses Church Post to Fight Evolution," May 19, 1923, 11.

33. Edward Larson, *Summer for the Gods: The Scopes Trial and America's Continuous Debate over Science and Religion* (New York: Basic Books, 1997), 22, 164, 156.

34. Clarence Darrow, *The Story of My Life* (New York: Charles Scribner's Sons, 1932), 279.

35. Darrow, *Clarence Darrow on Religion*, "The Pessimistic Versus the Optimistic View of Life," 8.

36. Ibid., "Why I Am an Agnostic," 43.

37. Kazin, *Godly Hero*, 272–274.

38. William Jennings Bryan, "Brother or Brute?" *Commoner* 20:11 (1920), 11.

39. Lienesch, *In the Beginning*, 72, 112.

40. Kazin, *Godly Hero*, 63.

41. *Boston Daily Globe*, "Attack of Apoplexy," July 27, 1925, 11.

42. Kazin, *Godly Hero*, 280.

43. Ibid., 275.

44. *Chicago Daily Tribune*, "Darrow Asks W. J. Bryan to Answer These," July 4, 1923, 1.

45. *Chicago Daily Tribune*, "Bryan Brushes Darrow Bible Queries Aside," July 5, 1923, 15.

46. Darrow, *Story of My Life*, 256.

47. Ibid., 244–245.

48. Ibid., 249–250.

49. Bryan, *Seven Questions*, 23. See Lienesch, *In the Beginning*, 64–67, on Bryan's campaign against colleges and universities.

50. Ronald L. Numbers, *The Creationists: From Scientific Creationism to Intelligent Design* (Cambridge, Mass.: Harvard University Press, 2006), 274–277.

51. Ibid., 71.

52. Ibid., 276.

53. The best account of the Scopes trial is Larson, *Summer for the Gods*. See also Ray Ginger, *Six Days or Forever? Tennessee v. John Thomas Scopes* (Boston: Beacon Press Books, 1958).

54. George M. Marsden, *Fundamentalism and American Culture: The Shaping of Twentieth-Century Evangelicalism: 1870–1925* (New York: Oxford University Press, 1980), 134.

55. Kazin, *Godly Hero*, 286.

56. H. L. Mencken, *A Religious Orgy in Tennessee: A Reporter's Account of the Scopes Monkey Trial* (Hoboken, N.J.: Melville House Publishing, undated), 131. It is a reprint of the articles originally published in 1925.

57. Ibid., 134.

58. Darrow, *Clarence Darrow on Religion*, "Why I Am an Agnostic," 48–49.

59. Leslie H. Allen, ed., *Bryan and Darrow at Dayton: The Record and Documents of the "Bible-Evolution Trial"* (New York: Russell and Russell, 1967), 138–154.

60. On Bryan's broad appeal, see Lienesch, *In the Beginning*, 140–152.

61. Kazin, *Godly Hero*, 173.

62. I know, Grant Wood's painting was five years later, in 1930.

63. Weinberg, *Clarence Darrow*, 351.

64. William Jennings Bryan, *Fighting to the Death*, "William Jennings Bryan's Last Message" (Dayton, Tenn.: William Jennings Bryan University, 1949), 1–2.

65. Ibid., 9.

66. Ibid., 3.

67. Ibid., 16.

68. *The World's Most Famous Court Trial: State of Tennessee v. John Thomas Scopes* (New York: Da Capo Press, 1971 [reprint of 1925 trial transcript]), 82, 87.

69. Ibid., 77–78.

70. Ibid., 84.

71. Larson, *Summer for the Gods*, 164.

72. Marsden, *Fundamentalism and American Culture*, 55–59.

73. Ibid., 20, 215.

Chapter 3. From the Scopes Trial to *Darwin on Trial*

1. The remark by Brady is from Act 1. See Jerome Lawrence and Robert E. Lee, *The Complete Text of Inherit the Wind* (New York: Random House, 1955), 41.

2. George M Marsden, *Fundamentalism and American Culture*, new edition (New York: Oxford University Press, 2006,) 189.

3. Walter Lippman, *A Preface to Morals* (New York: The Macmillan Company, 1929), 31, 60–61; Ronald Steel, *Walter Lippmann and the American Century* (Boston: Little, Brown and Company, 1980), 217–218.

4. Marsden, *Fundamentalism and American Culture*, 191.

5. On the impact of Mencken and the expanding meaning of *fundamentalism* and *obscurantist* labels, see Marsden, *Fundamentalism and American Culture*, 188.

6. Michael Lienesch, *In the Beginning: Fundamentalism, the Scopes Trial, and the Making of the Antievolution Movement* (Chapel Hill: University of North Carolina Press, 2007), 203–204.

7. On the restructuring of fundamentalism during the period, see Ronald L. Numbers, *The Creationists: From Scientific Creationism to Intelligent Design*, expanded edition (Cambridge, Mass.: Harvard University Press, 2006), 120–121; Ronald L.

Numbers, "Creationism in 20th Century America," *Science* 218:4572 (November 5, 1982), 540–541.

8. Karen Armstrong, *The Battle for God* (New York: Alfred A. Knopf, 2000), 216.

9. Numbers, "Creationism," 540–541; Lienesch, *In the Beginning*, 176, 203–204.

10. On the origin of the term *creationism*, see Ronald L. Numbers, *Darwinism Comes to America* (Cambridge, Mass.: Harvard University Press, 1998), 52–54. On Price and Bryan, see Randall J. Stephens and Karl W. Giberson, *The Anointed: Evangelical Truth in a Secular Age* (Cambridge, Mass.: The Belknap Press, 2011), 28–29.

11. Lienesch, *In the Beginning*, 193–196.

12. David Sehat, *The Myth of American Religious Freedom* (New York: Oxford University Press, 2011), 227–332; Armstrong, *Battle for God*, 182.

13. Lienesch, *In the Beginning*, 201–202; Tona J. Hangen, *Redeeming the Dial: Radio, Religion, & Popular Culture in America* (Chapel Hill: University of North Carolina Press, 2002), 125.

14. Lienesch, *In the Beginning*, 201–202. On the growth of radio ministries, see Hangen, *Redeeming the Dial*.

15. Armstrong, *Battle for God*, 214; Marsden, *Fundamentalism and American Culture*, 194.

16. Armstrong, *Battle for God*, 217.

17. John V. Grabiner and Peter D. Miller, "Effects of the Scopes Trial," *Science* 85:4154 (September 6, 1974): 832–837; Numbers, "Creationism," 540.

18. Sydney E. Ahlstrom, *A Religious History of the American People* (New Haven, Conn.: Yale University Press, 1972), 952; Martin E. Marty, *The New Shape of American Religion* (New York: Harper and Row, 1959), 15; Garry Wills, *Head and Heart: American Christianities* (New York: Penguin Press, 2007), 452–453. Some degree of error is unavoidable in any such survey. In this case, question wording would inevitably introduce error, particularly in light of the topic.

19. Marty, *New Shape of American Religion*, 12, 24.

20. Armstrong, *Battle for God*, 266–268, 275–277.

21. Sehat, *Myth of American Religious Freedom*, 227–232.

22. Ibid., 252–255.

23. Winthrop S. Hudson, *Religion in America* (New York: Charles Scribner's Sons, 1965), 384; Ahlstrom, *Religious History*, 957–958.

24. Wills, *Head and Heart*, 455–456.

25. *Time*, "The New Pictures," October 17, 1960, 97.

26. *Newsweek*, "New Films: The Monkey Trial," October 17, 1960, 114–117.

27. Numbers, "Creationism," 541.

28. Numbers, *Creationists*, 213.

29. Michael Ruse, *The Evolution-Creation Struggle* (Cambridge, Mass.: Harvard University Press, 2005), 240–242; Edward J. Larson, *Trial and Error: The American Controversy over Creation and Evolution* (New York: Oxford University Press, 1985), 58–92.

30. Eugenie C. Scott, *Evolution vs. Creationism: An Introduction* (Berkeley: University of California Press, 2004), 98–99; Numbers, "Creationism," 542–543.

31. *State v. Epperson*, 24 Ark. 922, 416 S.W. 2d, 322 (1967). *Epperson v. Arkansas*, 393 U.S. 97 (1968). On the history of court cases involving the teaching of evolution in public schools, see also Larson, *Trial and Error*, 107.

32. Marcel C. LaFollette, ed., *Creationism, Science, and the Law: The Arkansas Case*, (Cambridge, Mass.: The MIT Press, 1983), 5–6; Larson, *Trial and Error*, 119.

33. *Chicago Tribune*, "Shades of the Scopes Monkey Trial," November 17, 1968, 24.

34. *Time*, "The Supreme Court: Making Darwin Legal," November 22, 1968, 41.

35. *New York Times*, "Law: Children, Court Says, Should Be Seen and Heard," November 17, 1968, E5.

36. *New York Times*, "Court Ends Arkansas Darwinism Fan," November 13, 1968, 1.

37. *Newsweek*, "Supreme Court: Monkey Trial, 1968," November 25, 1968, 36–37. The remark about the "challenge" is erroneous because Scopes admitted teaching evolution, was a reluctant participant in the publicity farce, and even said there was no conflict between evolution and the Bible.

38. Dorothy Nelkin, *The Creation Controversy: Science or Scripture in the Schools* (New York: W. W. Norton and Company, 1982), 174.

39. Lienesch, *In the Beginning*, 207–209.

40. *Lemon v. Kurtzman*, 403 U.S. 602 (1971). See also, LaFollotte, *Creationism, Science, and the Law*, 7.

41. Scott, *Evolution vs. Creationism*, 193.

42. *New York Times*, "Parochial: Trouble For the Catholic Schools," July 4, 1971, E7.

43. *Los Angeles Times*, "Nixon Must Make Good on His Pledge to Aid Parochial Schools," July 12, 1971, D7.

44. Quoted in Larson, *Trial and Error*, 134–135.

45. Larson, *Trial and Error*, 139; LaFollette, *Creationism, Science, and the Law*, 6–7; Lienesch, *In the Beginning*, 208–209.

46. Larry A. Witham, *Where Darwin Meets the Bible* (New York: Oxford University Press, 2002), 116.

47. Armstrong, *Battle for God*, 309–314.

48. Quoted in Numbers, "Creationism," 543.

49. Numbers, *Creationists*, 330–331.

50. http://www.gallup.com/poll/21814, accessed October 2010. *McLean v. Arkansas Board of Education*. 529 F. Supp. 1255, 50 U.S. Law Week 2412 (1982).

51. LaFollette, *Creationism, Science, and the Law*, 1–2, 9–10. Overton's opinion is reprinted on 45–73.

52. *McLean v. Arkansas Board of Education*, 529 F. Supp. 1255, 50 U.S. Law Week 2412 (1982). See also, Lienesch, *In the Beginning*, 211; La Follette, *Creationism, Science, and the Law*, "Creationism in the News: Mass Media Coverage of the Arkansas Trial," 194.

53. LaFollette, *Creationism, Science, and the Law*, 193–198.

54. *Edwards v. Aguillard*, 482 U.S. 578 (1987).

55. Ibid. On the tortuous path of the case, see Raymond A. Eve and Francis B. Harrold, *The Creationist Movement in Modern America* (Boston: Twayne Publishers, 1990), 152–153; Larson, *Trial and Error*, 163–167.

56. *New York Times*: Stephen Jay Gould, "Evolution Wins Again"; Clarence Darrow, "Evolution, A Crime"; H. L. Mencken, "Scopes: Infidel"; January 12, 1982, A15.

57. *New York Times*, "Ruling Is Latest Strategy in Fighting Idea of Evolution," June 20, 1987, 1.

58. *Washington Post*, "Supreme Court Voids Creationism Law; 7–2 Ruling Deals Blow to Fundamentalists," June 20, 1987, A1.

59. Scott, *Evolution vs. Creationism*, 116.

60. On the early days of intelligent design, see Numbers, *Creationists*, 373–376.

61. Ibid., 373–374. The "Rottweiler" designation may have been inspired by the nickname of an earlier, and more prominent, advocate and ally of Darwin: Thomas Henry Huxley (1825–1895), who was known as "Darwin's Bulldog."

62. www.gallup.com/poll/21814, accessed October 2010.

Chapter 4. Intelligent Design and Resurgent Creationism

1. Ronald L. Numbers, *The Creationists: From Scientific Creationism to Intelligent Design*, expanded edition (Cambridge, Mass.: Harvard University Press, 2006), 353–355, 368–370. The Dawkins remark in the epigraph is from *New York Times Book Review*, "Richard Dawkins Review of Blueprints: Solving the Mystery of Evolution," April 9, 1989, 34. The Johnson remark is from "Afterword: How to Sink a Battleship," in William Dembski, ed., *Mere Creation: Science, Faith & Intelligent Design* (Downers Grove, Ill.: Inter-Varsity Press, 1998), 451.

2. http://www.gallup.com/poll/21814/Evolution-Creationism-Intelligent-Design, accessed October 29, 2012.

3. Numbers, *Creationists*, 353–354. On the American Scientific Affiliation, see http://www.asa3.org, accessed February 23, 2013.

4. Barbara Forrest and Paul R. Gross, *Creationism's Trojan Horse: The Wedge of Intelligent Design* (New York: Oxford University Press, 2004), 116.

5. Charles B. Thaxton, Walter L. Bradley, Roger L. Olsen, *The Mystery of Life's Origin: Reassessing Current Theories* (New York: Philosophical Library, 1984), vi–vii.

6. Ibid., 188–214.

7. Peter Applebome, "70 Years after Scopes Trial, Creation Debate Lives," *New York Times*, March 10, 1996, 1.

8. Numbers, *Creationists*, 373–396. On the differences of creationists and intelligent design, also see Eugenie C. Scott, *Evolution vs. Creationism, An Introduction* (Los Angeles: University of California Press, 2004), 128.

9. Phillip E. Johnson, *Darwin on Trial* (Washington, D.C.: Regnery Gateway, 1991), 4, footnote 1.

10. Ibid., 4–6, 8.

11. William Dembski, ed., *Darwin's Nemesis: Phillip Johnson and the Intelligent Design Movement* (Leicester, England: Inter-Varsity Press, 2006), 12–13, 26–27, 90, 100, 308–309.

12. Michael Ruse, *The Evolution-Creation Struggle* (Cambridge, Mass.: Harvard University Press, 2005), 250–252.

13. Michael J. Behe, *Darwin's Black Box: The Biochemical Challenge to Evolution* (New York: The Free Press, 1996), 192–193.

14. Ibid., 36–39, 46–47, 250–251.

15. Ibid., 250.

16. Daniel C. Dennett, *Darwin's Dangerous Idea: Evolution and the Meanings of Life* (New York: Simon and Schuster, 1995), 515–516.

17. Numbers, *Creationists*, 383–384.

18. And I immediately concede the point that these often are the same media that believe "news" includes celebrity gossip and horoscopes.

19. Richard Vara, "Professor Says Evolution Lacks Scientific Evidence," *Houston Chronicle*, June 22, 1991, 1. Vara is the newspaper's religion writer.

20. *Colorado Springs Gazette Telegraph*, "Analysis of Darwinism Earns Author Label as 'Intellectual Caveman,'" February 1, 1992, D3.

21. Richard Saltus, "Educator Reports on Creationism," *Boston Globe*, February 14, 1993, 36.

22. Jack Hitt, "On Earth as It Is in Heaven," *Harper's Magazine*, November 1, 1996 (vol.1, issue 1758), 51.

23. Boyce Rensberger, "How Science Responds when Creationists Criticize Evolution," *Washington Post*, January 8, 1997, H1.

24. Jeffrey Weiss, "The Origin of Speeches [*sic*] Law Professor Builds His Case against Darwinism," *Dallas Morning News*, August 2, 1997, 1G.

25. *Atlanta Journal-Constitution*, "UGA Scholars Battle Idea of Faithless Academics," December 21, 1997, H10.

26. Laurie Goldstein, "Week in Review: New Light for Creationism," *New York Times*, December 21, 1997, 1.

27. PBS, *The Firing Line*, "Creation-Evolution Debate: 'Resolved: The Evolutionists Should Acknowledge Creation,'" December 4, 1997, www.bringyou.to/apologetics/p45.htm, accessed October 25, 2012.

28. PBS, *NOVA*, "Defending Intelligent Design," www.pbs.org/wgbh/nova/evolution/defense-intelligent-design.html, accessed October 25, 2012.

29. Stephen Jay Gould, *Rocks of Ages: Science and Religion in the Fullness of Life* (New York: Ballantine Books, 1999), 147.

30. *Edwards v. Aguillard*, 482 U.S. 578 (1987).

31. Ibid.

32. *Webster v. New Lenox School District No. 122*, 482 U.S. 578 (1987), Seventh Circuit Court of Appeals (1990).

33. *Peloza v. Capistrano Unified School District*, 37 F.3d 517, Ninth Circuit (1994).

34. *Peloza v. Capistrano Unified School District*, 515 U.S. 1173 (1995).

35. *Hellend v. South Bend Community School Corporation*, 93 F.3d 327, 329, Seventh Circuit (1996), cert. denied, 519 U.S. 1092 (1997).

36. *Frieler v. Tangipahoa Parish Board of Education*, 185 F. 3d 337, 341, 346; Fifth Circuit 201 F. 3d 602 (2000); *Tangipahoa Parish Board of Education et al. v. Herb Freiler et al.*, cert. denied, No. 99-1625, 530 U.S. 1251 (2000).

37. *LeVake v. Independent School District no. 656*, 625 N.W.2d 502, 505–506, 509, Minnesota Court of Appeal (2000), cert. denied, 534 U.S. 1081 (2002).

38. *Moeller v. Schrenko*, 554 S.E.2d 200, Georgia Court of Appeals (2001).

39. Ibid., 200.

40. *Selman et al. v. Cobb County School District*, no. 1: 02CV2325-CC, N.D. Georgia, 390 F. Supp. 2d 1286 (2005), 1312, 1302, 1306, 1292.

41. Ibid., 1310.

42. Susan Jacoby, *Freethinkers: A History of American Secularism* (New York: Henry Holt and Company, 2004), 131–134.

43. Johnson, *Darwin on Trial*, 9.

44. Numerous articles by Johnson and others at the web site of the Bible Creation Society press the point of creationist tolerance versus mainstream science's exclusivity. The society home page dedicates a whole "file" to Richard Dawkins. See http://www.biblicalcreation.org.uk/, accessed June 22, 2011.

45. PBS, "Interview at PBS with Bill Moyers," December 3, 2004, old.richarddawkins.net/videos/64-interview-at-PBS-with-bill-moyers, accessed October 24, 2012.

46. PBS, *Think Tank*, "Talking about Evolution with Richard Dawkins," October 18, 2001, http://www.pbs.org/thinktank/transcript410.html, accessed October 24, 2012.

47. Dan Cray, "God vs. Science," *Time*, November 5, 2006.

48. Frank S. Ravitch, *Marketing Intelligent Design: Law and the Creationist Agenda* (New York: Cambridge University Press, 2011); see 148–155 for a more in-depth discussion of audiences and the marketing of Dawkins.

49. Dembski, *Darwin's Nemesis*, 90.

50. Gould, *Rocks of Ages*, 17.

51. Richard Dawkins, *The God Delusion* (New York: Houghton Mifflin Company, 2006), 54–61.

52. David van Biema, Alice Park, Dan Cray, Jeff Israely, and David Bjerklie, "God vs. Science," *Time*, November 13, 2006, 48–55.

53. Jacoby, *Freethinkers*, 131.

54. Henry Schaefer in Dembski, ed., *Mere Creation*, Foreword, 11; Forrest and Gross, *Creationism's Trojan Horse*, 20–22.

55. Forrest and Gross, *Creationism's Trojan Horse*, 20–22.

56. Bruce Chapman in Dembski, ed., *Mere Creation*, Postscript, 458.

57. Dembski, ed., *Mere Creation*, Introduction, 13–14.

58. Ibid., 19.

59. Ibid., 23; Walter L. Bradley in Dembski, ed., *Mere Creation*, "Chapter 1: Nature," 33.

60. Johnson, "Afterword," in Dembski, ed., *Mere Creation*, 451.

61. Ibid., 453.

62. See the web site for the National Center for Science Education for a pdf file of the original *Wedge Document*: http://ncse.com/webfm_send/747. A number of documents now exist on the web purporting to be the "original" web document, but some are suspect because the web site sponsors include a variety of creationist groups, including the Discovery Institute. These newer versions may have been altered from the original one produced at the Biola University conference.

63. Ibid.

Chapter 5. Science on Trial

1. The epigraph is from *The World's Most Famous Court Trial: State of Tennessee v. John Thomas Scopes* (New York: Da Capo Press, 1971 [reprint of 1925 edition of trial transcript]), 299.

2. *New York Times*, "Now Arguing Near You: The Evolution Drama," October 12, 2005, E4.

3. *Washington Post*, "Evolution Debate in Kansas Spurs Battle over School Materials," October 28, 2005, A2.

4. "Evolution's Grass-Roots Defender Grows in Va.," *Washington Post*, July 20, 2005, B1. The *Los Angeles Times* quoted Michael Ruse, author of *Evolution-Creation Struggle*, as calling the teaching of intelligent design "an especially American issue—a flashpoint in a national culture war." The article did not give details on other aspects of the war, only that intelligent design was central to the conflict; *Los Angeles Times*, "Exhibit Explores the Evolution of Darwin's Ideas and Research," December 4, 2005, A20.

5. *Harrisburg Patriot-News*, "Intelligent Design Trial Attracts Global Curiosity," October 4, 2005, A14; "Education Could Use Intelligence in Its Design," October 2, 2005, F1. A good narrative account of the case is Laura Lebo, *The Devil in Dover: An Insider's Story of Dogma v. Darwin in Small-Town America* (New York: The New Press, 2008).

6. Edward Humes, *Monkey Girl: Evolution, Education, Religion, and the Battle for America's Soul* (New York: Harper Perennial, 2007), 176. The director is Eugenie Scott, quoted on 177.

7. Humes, *Monkey Girl*, 64, 75–76. Barbara Forrest and Paul R. Gross, *Creationism's Trojan Horse: The Wedge of Intelligent Design* (New York: Oxford University Press, 2004) is an in-depth account of the *Wedge Document*, its origins, creators, and content. On the document's verification, see especially 26–33.

8. Humes, *Monkey Girl*, 158–178, is a detailed account of the Kansas hearings.

9. Matthew Chapman, *40 Days and 40 Nights* (New York: HarperCollins, 2008), 11.

10. *Tammy Kitzmiller, et al. v. Dover Area School District, et al.*, 400 F. Supp. 2d 707 (2005). See also Humes, *Monkey Girl*, 201–203, for details leading up to the lawsuit.

11. Forrest and Gross, *Creationism's Trojan Horse*, 325–326.

12. *Tammy Kitzmiller, et al. v. Dover Area School District, et al.*, 400 F. Supp. 2d 707, 32–34 (2005). See also Forrest and Gross, *Creationism's Trojan Horse*, 329–330.

13. *Tammy Kitzmiller, et al. v. Dover Area School District, et al.*, 400 F. Supp. 2d 707, 32–34 (2005), 74–80. On Jones's criticism of Behe, see also Forrest and Gross, *Creationism's Trojan Horse*, 331.

14. *Tammy Kitzmiller, et al. v. Dover Area School District, et al.*, 400 F. Supp. 2d 707, 32–34 (2005), 26. In Lebec, California, a group of parents sued the school district over a new course called "Philosophy of Design." School officials were careful to point out that the course was not offered in a science curriculum. Parents who filed the suit charged that the course was designed to get around Judge John Jones's decision in Dover. See *New York Times*, "California Parents File Suit over Origins of Life Course," January 11, 2006, A1.

15. *Los Angeles Times*, "Judge Says 'Intelligent Design' Is Not Science," December 21, 2005, 1.

16. *New York Times*, "Issuing Rebuke, Judge Rejects Teaching of Intelligent Design," December 21, 2005, 1; *Washington Post*, "Judge Rules against 'Intelligent Design'; Dover, Pa., Can't Teach Evolution Alternative," December 21, 2005, A1; *Washington Post*, "Santorum Breaks with Christian-Rights Law Center," December 23, 2005, A11.

17. National Public Radio, "Interview: Barbara Forrest Discusses Pennsylvania Trial Sending Intelligent Design out of Class," December 23, 2005.

18. Humes, *Monkey Girl*, 314–315, 333.

19. *ABC World News Tonight*, January 18, 2005; August 11, 2005; see also September 26, 2005.

20. *CBS Evening News*, September 26, 2005; May 12, 2005.

21. *NBC Nightly News*, November 8, 2005; August 10, 2005.

22. *ABC Nightline*, January 13, 2005.

23. CNN *Crossfire*, May 5, 2005.

24. *ABC Nightline*, August 10, 2005. Ronald Numbers, of the University of Wisconsin–Madison and author of *Creationists*, called it "cheating" to simply stop scientific inquiry by declaring something the handiwork of God. The document surfaced on the internet in 1999, and outlines the plan from the Center for Renewal of Culture.

25. *Kansas City Star*, "Kansas Case of Déjà vu: Evolution Critics Win," November 9, 2005, A1.

26. *New York Times*, "Sleepy Election Is Jolted by Evolution," May 17, 2005, A12.

27. *Washington Post*, "Advocates of 'Intelligent Design' Vow to Continue Despite Ruling," December 22, 2005, A28.

28. *Washington Post*, "Judge Rules against 'Intelligent Design,'" December 25, 2005, A3.

29. *Atlanta Journal-Constitution*, "Dover, Pa., Is Still God's Country," November 14, 2005, A12.

30. *Harrisburg Patriot-News*, "Creation Fight Evolves beyond Dover," February 13, 2005, A1. Many other stories shared this theme, including "Dover Dilemma Speakers Favor Science," January 26, 2005, B1; "Unnecessary Conflict Does Harm to Religion," February 25, 2005, A11; "Lawyer Asks to Join Dover Suit," March 1, 2005, B1; "Bills Would Shield Schools from Church-State Dispute," April 9, 2005, A1.

31. *Harrisburg Patriot-News*, "Dover to Rerun School Election," December 30, 2005, 4.

32. *NBC Nightly News*, May 9, 2005.

33. *Dateline NBC*, October 28, 2005; *NBC News*, May 9, 2005.

34. *CBS Evening News*, December 20, 2005.

35. *CBS Evening News*, September 26, 2005.

36. Michael D. Lemonick, Noah Isackson, Jeffry Ressner, "Stealth Attack on Evolution: Who Is behind the Movement to Give Equal Time to Darwin's Critics, and What Do They Really Want?" *Time*, January 31, 2005, 53; Claudia Wallis, "The Evolution Wars," *Time*, August 15, 2005, 26.

37. Eric Cornell, "What Was God Thinking? Science Can't Tell," *Time*, November 14, 2005, 100; Michael Lemonick, "Much Ado about Evolution," *Time*, November 21, 2005, 23.

38. Charles Krauthammer, "Let's Have No More Monkey Trials," *Time*, August 8, 2005, 78.

39. Wallis, "The Evolution Wars," 26.

40. Jerry Adler, "Doubting Darwin," *Newsweek*, February 7, 2005, 50, 57; Jonathan Alter, "Monkey See, Monkey Do," *Newsweek*, August 15, 2005, 27.

41. Adler, "Doubting Darwin," 44.

42. Jonathan Alter, "Monkey See, Monkey Do," *Newsweek*, August 15, 2005, 27.

43. William Lee Adams, "The Classroom: Other Schools of Thought," *Newsweek*, November 28, 2005, 57. An article by Jerry Adler, "Evolution of a Scientist," *Newsweek*, November 28, 2005, 50, noted the story of Darwin's sudden conversion to Christianity on his deathbed and his rejection of evolution. On the recantation myth and Darwin's agnosticism, see Edward Caudill, *Darwinian Myths: The Legends and Misuses of a Theory* (Knoxville: University of Tennessee Press, 1997), 46–63.

44. Gaye Tuchman, *News: A Study in the Construction of Reality* (New York: The Free Press, 1978), 82–86, 89–93, 166–167, 190. Tuchman points to an Enlightenment-era model of public debate and discourse as a means of assessing ideas.

45. Chris Mooney and Matthew C. Nisbet, "Undoing Darwin," *Columbia Journalism Review* (Fall 2005), 34.

46. Ibid., 35–36.

47. Ibid, 32, 36.

48. Roger Streitmatter, *Mightier Than the Sword: How the News Media Have Shaped American History*, 2nd ed. (Philadelphia: Westview Press, 2008), 259–260.

Chapter 6. Into the Mainstream

1. The epigraph is from Clarence Darrow, *The Story of My Life* (New York: Charles Scribner's Sons, 1960), 91.

2. ncse.com/rncse/21/1–2/teaching-evolution-do-state-science-standards-matter, accessed October 10, 2011.

3. *Creationism in 2001: A State-by-State Report*, commissioned by People for the American Way, http://law2.umkc.edu/faculty/projects/ftrials/conlaw/creationism report.pdf, accessed August 30, 2011. For the poll results, see www.gallup.com/poll21814/evolution-creationism-intelligent-design.aspx, accessed October 10, 2011; pewresearch.org/pubs/1105/darwin-debate-religion-evolution, accessed October 17, 2011.

4. David Applegate, "Creationists Open New Front," www.agiweb.org/geotimes/july00/scenehtml, accessed October 10, 2011. The American Geological Institute is a federation of geoscientific associations.

5. *Congressional Record*, June 14, 2000, page H4480-H3382, from *The Congressional Record Online* via GPO Access, wais.access.gpo.gov. *Creationism in 2001: A State-by-State Report*, http://law2.umkc.edu/faculty/projects/ftrials/conlaw/creationismreport.pdf, accessed August 30, 2011.

6. *Washington Post*, "Battle on Teaching Evolution Sharpens," March 14, 2005, A1.

7. Sen. Rick Santorum, "Illiberal Education in Ohio Schools," *Washington Times*, March 14, 2002, A14.

8. Sen. Rick Santorum, "It Takes a Family," National Public Radio, broadcast August 4, 2005, http://www.npr.org/templates/story/story.php?storyId=4784905, accessed August 30, 2011.

9. *Boston Globe*, "The Religious Right Faces Its Purgatory," January 10, 2006, A3.

10. William A. Dembski, ed., *Darwin's Nemesis: Phillip Johnson and the Intelligent Design Movement*, Foreword by Sen. Rick Santorum (Leicester, England: Inter-Varsity Press, 2006), 9–11.

11. Richard G. Hutcheson Jr., *God in the White House: How Religion Has Changed the Modern Presidency* (New York: Macmillan Publishing Company, 1988), 153–155.

12. Chris Mooney, *The Republican War on Science* (New York: Basic Books, 2005), 36–37.

13. "Republican Candidate Picks Fight with Darwin," *Science*, 29:12 (September 1980), 1214.

14. www.americanrhetoric.org/speeches/ronaldreagan, accessed September 9, 2011.

15. Quoted in the *New York Times*, "In Their Own Words," February 19, 1996, A10.

16. *New York Times*, "Equal Time for Nonsense," July 29, 1996, A19.

17. *New York Times*, "Media Martyr," March 3, 1996, 15, E15.

18. Mooney, *Republican War on Science*, 9; *New York Times*, "At G.O.P. Debate, Candidates Played to Conservatives," May 5, 2007, A10. In the PBS *Frontline* documentary, *The Jesus Factor*, George Bush was unequivocal about the influence of evangelical Christianity on his politics as well as his personal life. The documentary aired April 29, 2004. See also *New York Times*, "Understanding the President and His God," April 22, 2004, A1.

19. A tape of the question at the debate and the responses is www.youtube.com/watch?v=t4c8t3zdE, accessed October 3, 2011. The positions of the candidates are given in more detail in *New York Times*, "At G.O.P. Debate, Candidates Played to Conservatives," May 5, 2007, A10; www.msnbc.msn,com/id/22111924/ns/politics-decision 08, accessed October 3, 2011; Sam Brownback, *New York Times*, "What I Think about Evolution," May 31, 2007, A19.

20. www.gallup.com/poll/27847, accessed September 14, 2011.

21. Todd Charles Wood, "No Room for the Creator?" *Answers* 4:1 (January–March 2009), 12–13.

22. *Washington Post*, "Perry Hits Bump on Campaign Trail in N.H.," August 18, 2011, A2; *USA Today*, "Perry Stances Hurt GOP, Huntsman Says," August 22, 2011, 5A; *USA Today*, "Rick Perry: Evolution Is a 'Theory' with 'Gaps,'" August 18, 2011; www.outside-the-beltway.com/michele-bachman-schools-should-teach-intelligent-design/, accessed October 4, 2011; www.huffingtonpost.com/2011/09/19/jon-huntsman-rick-perry-evolution, accessed October 4, 2011; National Public Radio, "Interview: Chris Mooney Discusses 'The Republican War on Science,'" November 11, 2005.

23. www.gallup.com/poll/108226 and www.gallup.com/poll/121814, accessed September 14, 2011. The designation "the past quarter-century" refers to how far back the Gallup questions go on the subject. The number of young-Earth adherents probably has been an even greater proportion of the unpolled past.

24. www.gallup.com/poll/16462, accessed September 14, 2011.

25. www.gallup.com/poll/19207, accessed September 14, 2011.

26. www.gallup.com/poll/21811, accessed September 14, 2011.

27. www.gallup.com/poll/16462, accessed September 14, 2011.

28. www.gallup.com/poll/114544, accessed September 14, 2011.

29. www.gallup.com/poll/148427, accessed September 14, 2011.

30. *New York Times*, "Bush Remarks Roil Debate on Teaching Evolution," August 3, 2005, A14.

31. John G. West, associate director, Center for Science and Culture, Discovery Institute, "Intelligent Design Is Sorely Misunderstood," *Seattle Post-Intelligencer*, August 9, 2005, B7; *Christian Century*, "Bush Endorses Teaching of 'Intelligent Design,'" August 23, 2005, 12.

32. *New York Times*, "Bush Remarks Roil Debate on Teaching of Evolution," August 3, 2005, A14. On the Discovery Institute support for the president's position, see www.discovery.org/a/2764, accessed September 14, 2011.

33. *New York Times*, "Politicized Scholars Put Evolution on the Defensive," August 21, 2005, A1.

34. *Time*, "The Evolution Wars," August 15, 2005, 27–35.

35. www.watchcartoononline.com/the-simpsons-season-17-episode-21-the-monkey -suit, accessed September 27, 2011.

36. www.answersingenesis.org/articles/2006/05/15/Simpsons-satire-special-creation, accessed August 25, 2011.

37. PBS, *NOVA*, "Judgment Day: Intelligent Design on Trial," WGBH Educational Foundation and Vulcan Productions, Boston, aired November 13, 2007.

38. www.answersingenesis.org/articles/2007/11/14/over-after-dover, accessed September 28, 2011.

39. www.discovery.org/a/4300/, accessed September 28, 2011.

40. Randy Olson, *Flock of Dodos: The Evolution-Intelligent Design Circus* (Los Angeles: Prairie Starfish Productions, 2006).

41. www.discovery.org/a/4052, accessed September 28, 2011. For a good critique of Wells and *Icons*, see Massimo Pigliucci, "No Icon of Evolution: A Review of 'Icons of Evolution,'" *Bioscience* 51:5 (May 2001), 411–414.

42. www.answersingenesis.org/article/2009/09/13/flocking-to-video-stores, accessed September 28, 2011.

43. Lawrence S. Lerner, *Good Science, Bad Science: Teaching Evolution in the States* (Dayton, Ohio: Thomas B. Fordham Foundation, 2000), http://www.texscience.org/ files/lerner-fordham, accessed August 31, 2011. An updated study, published in 2009, in *Evolution Education and Outreach* said evolution was covered more extensively in classrooms than a decade earlier, but certain creationist language became more common. In the NCSE study, eleven states had an unsatisfactory rating on evolution.

44. *Creationism in 2001: A State-by-State Report*, http://law2.umkc.edu/faculty/ projects/ftrials/conlaw/creationismreport.pdf, accessed August 30, 2011. accessed August 30, 2011.

45. Ibid.; www.Pandasthumb.org/archives/2006/03/ky-governor_kno_html, accessed September 16, 2011.

46. http://toddcwood.blogspot.com, accessed September 8, 2011.

47. www.faqs.org/tax-exempt/ky/answers-in-genesis-of-kentucky-inc.html, accessed September 16, 2011; www.bbb.org.charity-reviews/cincinnati/religious/answers -in-genesis-of-kentucky-in-petersburg-ky-8930, accessed September 16, 2011; www .faqs.org/tax-exempt/ca/national-center-for-science-education.html, accessed September 16, 2011.

48. http://lippard.blog.com; www.faqs.org/tax-exempt/tx/institute-for-creation- research.html, accessed September 16, 2011.

49. www.icr.org, accessed October 11, 2011. More information is available at http:// www.bing.com (accessed February 17, 2013), but it does not reveal the donors. It is the cite for "Charity Navigator: Your Guide to Intelligent Giving."

50. www.discovery.org/about.php, accessed October 11, 2011.

51. www.discovery.org/csc, accessed October 11, 2011. See also, Eugenie C. Scott, *Evolution vs. Creationism: An Introduction* (Los Angeles: University of California Press,

2004), 125; Ronald L. Numbers, *The Creationists: From Scientific Creationism to Intelligent Design*, expanded edition (Cambridge, Mass.: Harvard University Press, 2006), 381–383; *New York Times*, "Politicized Scholars Put Evolution on the Defensive," August 21, 2005, A1; www.faqs.org/tax-exempt/wa/discovery-institute.html, accessed September 16, 2011.

52. *New York Times*, "Politicized Scholars Put Evolution on the Defensive," August 21, 2005, 1. Along with a "well-tooled electoral campaign" with poll-tested messages and web logs, the institute accelerated the evolution battle, the *Times* said, with activities that included lobbying of the Kansas state board of education, showing a film at the Smithsonian that promoted intelligent design, and even helping a Roman Catholic cardinal place an opinion piece in the *Times* that tried to distance the church from evolution.

53. Ibid.

54. www.discovery.org/a/1537, accessed July 7, 2011.

55. Numbers, *Creationists*; see 180–184 on ASA background; see 201 on "scientific creationism" term. See also the ASA's web site at www.asa3.org, accessed February 17, 2013.

56. George E. Webb, *The Evolution Controversy in America* (Lexington: University Press of Kentucky, 1994), 157–158.

57. www.faqs.org/tax-exempt/ma/american-scientific-affiliation.html, accessed September 16, 2011.

58. On the Creation Research Society link to the American Scientific Affiliation, see Numbers, *Creationists*, 249–250, 255.

59. www.creationresearch.org/hisaims.htm, accessed October 11, 2011; www.faqs.org/tax-exempt/az/creation-research-society.html, accessed September 16, 2011.

Chapter 7. Creationism's Web

1. The epigraph is from *The World's Most Famous Court Trial: State of Tennessee v. John Thomas Scopes* (New York: Da Capo Press, 1971 [reprint of 1925 edition of trial transcript]), 287.

2. www.answersingenesis.org/articles/2000/07/24/the-scopes-trial-big-deal, accessed August 25, 2011; also from personal visit to the museum in July 2011.

3. http://www.icr.org/article/mr-bryan-on-evolution, accessed June 26, 2011. The other articles also are from www.icr.org, accessed June 26, 2011.

4. www.newcreation.net/museums.html, accessed October 12, 2011. The number of museums is fluid because many are ambitions of individuals or of small congregations.

5. *New York Times*, "Creationist Captain Sees Battle 'Hotting Up,'" December 1, 1999, 18.

6. Ken Ham, *The Lie: Evolution* (Green Forest, Ariz.: Master Books, 1987).

7. Randall J. Stephens and Karl W. Gilberson, *The Anointed: Evangelical Truth in a Secular Age* (Cambridge, Mass.: The Belknap Press, 2011), 42–43.

8. Ibid., 43–44.

9. www.answersingenesis.org/about/history, accessed August 23, 2011; Eugenie C. Scott, *Evolution vs. Creation: An Introduction* (Berkeley: University of California Press, 2004), 110.

10. *New York Times Magazine*, "Rock of Ages, Ages of Rock," November 25, 2007, 30.

11. *Raleigh News & Observer*, May 20, 2007, "Museum Puts Dinosaurs in Eden," D1.

12. *New York Times*, "Adam and Eve in the Land of the Dinosaurs," May 24, 2007, E1; see also *New York Times*, "Museums, God and Man," April 27, 2007, F2.

13. *Washington Post*, "A Monument to Creation: Kentucky Museum Discounts Centuries of Research, Critics Say," May 27, 2007, 3.

14. *Chicago Sun Times*, "Taking Aim at Evolution; Creation Museum Opens Monday; Literal View of Bible Riles Critics," May 27, 2007, A9.

15. *USA Today*, "Creation Museum Juxtaposes Dinosaurs, Noah's Ark," May 26, 2007; *USA Today*, "Ky. Creation Museum Opens to Thousands," May 28, 2007; *Newsweek*, "BeliefWatch: Edutainment," June 4, 2007, 10. The *USA Today* articles are available at www.usatoday.com, accessed November 23, 2011.

16. *Dayton Daily News*, "Young Earth Creationists Open Museum," May 26, 2007, 8.

17. *Dayton Daily News*, "Creationist Display Creating Controversy," May 26, 2007, A6.

18. *Dayton Daily News*, "According to Genesis, Displays at Northern Kentucky Venue Show a Strictly Biblical View of Earth," May 26, 2007, A1.

19. www.answersingenesis.org/article/nab2/why-is-scopes-trial-significant, accessed September 1, 2011.

20. Ibid.

21. A number of the museum displays are described in *Journey through the Creation Museum: Prepare to Believe* (Green Forest, Ariz.: Master Books, 2008).

22. Todd Charles Wood, "Natural Selection: Theory or Reality?" *Answers*, 4:1 (January–March 2009), 50, note 10.

23. *New York Times*, "In or Out of Eden, One Man's Unicorn May Be Another's Apatosaurus," June 4, 2007, E3.

24. *New York Times Magazine*, "Rock of Ages, Ages of Rock," November 25, 2007, 30.

25. CNN, *CNN Newsroom*, May 22, 2007; CNN, *The Situation Room*, May 28, 2007.

26. CNN, *Special Investigations*, May 19, 2007.

27. Ibid.

28. ABC News, *Good Morning America*, "Did God Create Dinosaurs?; New Museum Takes on Evolution," May 25, 2007; ABC News, *World News*, "Creation Debate; Museum Challenged Science on Creation," May 27, 2007.

29. The following *USA Today* articles—"Kentucky Creation Museum Draws Crowds along with Controversy," July 7, 2007; "Business Booming at Controversial

Creation Museum," August 2, 2007; "Creation Museum Surpasses Expectations," November 2, 2007—are available at www.usatoday.com/travel, accessed November 23, 2011. *New York Times*, "In Kentucky, Noah's Ark Theme Park Is Planned," www .nytimes.com/2010/12/06/us/06ark.html?pagewanted=print, accessed October 7, 2011; Nky.com, http://nky.cincinnati.com/article/AB/20101130;NEWS0103/311300043/ Beshear-to-announce-creationist-theme-park-for-kentucky, accessed October 7, 2011.

30. www.answersingenesis.org, accessed August 23, 2011.

31. www.answersingenesis.org/articles/2011/04/29/feedback-is-ark-encounter-a -waste-of-money, accessed October 6, 2011.

32. *Lexington Herald-Leader*, "$43 Million Tax Break Approved for Ark Encounter Theme Park," May 20, 2011, www.kentucky.com/2011/05/20/1745988, accessed November 23, 2011.

33. www.answersingenesis.org/articles/2010/12/10/feedback-taxpayers-not-paying -to-build-ark-encounter, accessed October 4, 2011.

34. www.answersingenesis.org/articles/2011/06/14/remarkable-charges-by-christian, accessed October 4, 2011.

35. www.answersingenesis.org/articles/au/clergyman-opposes-ark, accessed October 4, 2011.

36. www.answersingenesis.org/articles/2011/01/27/clergyman-opposes-ark-en-counter, accessed October 4, 2011.

37. http:arkencounter.com/blog/2011/09/26/its-time-to-check-in, accessed October 5, 2011.

38. www.answersingenesis.org/articles/2011/12/10/feedback-taxpayers-not-paying -to-build-ark-encounter, accessed October 4, 2011; www.answersingenesis.org/articles/ 2011/12/06/media-mania, accessed October 4, 2011.

39. www.discovery.org/a/2764, accessed October 13, 2011.

40. www.answersingenesis/articles/2005/08/04/president-bush-origins, accessed October 13, 2011.

41. www.answersingenesis.org/articles/2005/08/04/time-evolution-wars, accessed August 25, 2011.

42. www.answersingenesis.org/about/history, accessed August 23, 2011.

43. www.answersingenesis.org/articles/cm/v20/n2/antidote-to-superstition, accessed August 25, 2011.

44. www.answersingenesis.org/articles/cm/v19/n1/inherit-the-wind-analysis, accessed August 25, 2011.

45. The articles to which I allude are numerous and all are available at www .answersingenesis.org, accessed August 25, 2011. The most illustrative of these points are "Inherit the Wind: An Historical Analysis," by David Menton, December 1, 1996; "The Full 'Scope' of the Trial of the Century," July 20, 2000; "Monkeying with the Media: A Case Study of the Scopes Trial and the Media Impact," by Rick Barry, June 7, 2007; "The Scopes Trial . . . What's the Big Deal?" by Stacia Byers, July 24, 2000; "The Scopes 'Monkey Trial'—80 Years Later," by David Menton, July 11, 2005; "The Scopes

Trial Quiz: Deciphering Fact from Fiction," July 11, 2005; "An 'Expected' Slant?" by David Menton, April 11, 2006; "Amen to 'BreakPoint,'" by Ken Ham, October 27, 2005; "A Shocking Movie List?" by Mark Looy, June 16, 2006; "The Scopes Retrial," June 24, 2010.

46. www.answersingenesis.org/articles/2000/07/20/full-scope, accessed August 25, 2011.

47. PBS, *NOVA*, "Evolution" (Boston: WBGH Educational Foundation, Clear Blue Sky Productions, 2001).

48. *Getting the Facts Straight: A Viewer's Guide to PBS's Evolution* (Seattle: Discovery Institute Press, 2001), 9. This also is available at www.arn.org/pbsevolution/vguide .pdf, accessed September 20, 2011.

49. Ibid., 14, 15.

50. Ibid., 15.

51. Stephen Gould, Introduction, in Bjorn Kurten, *Dance of the Tiger: A Novel of the Ice Age* (New York: Pantheon Books, 1980), xvii.

52. Discovery Institute, *Getting the Facts Straight*, 141.

53. Ibid., 141–143.

54. Ibid., 107, 145–146.

55. Ibid., 149–151.

56. William Jennings Bryan, *Fighting to Death for the Bible*, "William Jennings Bryan's Last Message" (Dayton, Tenn.: William Jennings Bryan University, 1949), 5.

57. www.answersingenesis.org/articles/2009/11/13/unbecoming-of-humans; www .answersingenesis.org/articles/2009/11/07/news-to-note; www.answersingenesis.org/ articles/2009/11/10/NOVA-no-viable-answers, all accessed October 3, 2011.

58. On the eugenics movement in America, see Daniel J. Kevles, *In the Name of Eugenics: Genetics and the Uses of Human Heredity* (Berkeley: University of California Press, 1985).

59. *Variety*, "Image Launches Slingshot Pictures," September 27, 2011, www.variety.com/ article/VR1118043530?, accessed November 22, 2011; *Flint Journal*, "'Alleged' Screenwriter Fred Foote Discusses the Story behind His Story on the Scopes Monkey Trial," September 24, 2008, www.mlive.com/news/flint/index.ssf/2009/09/alleged_screenwriter _fred_foote.html, accessed November 22, 2011; *Business Wire*, "Image Entertainment to Launch New Faith-Based Label Slingshot Pictures," September 28, 2011, www.business wire.com/news/home/20110928006364/en/Image-Entertainment-Faith-Based-Label -Slingshot-Pictures, accessed November 22, 2011.

60. Personal visit to museum, July 2011.

61. Wood, "Natural Selection," 49.

62. Personal visit, July 2011.

63. On the *Buck v. Bell* case, see Kevles, *In the Name of Eugenics*, 110–111. The politicizing of genetics into eugenics in the early twentieth century is discussed in Edward Caudill, *Darwinian Myths: The Legends and Misuses of the Theory*, Chapter 6, "Eugenics: The Political Science" (Knoxville: University of Tennessee Press, 1997).

64. *Charles Darwin: His Life and Impact* (Petersburg, Ky.: Answers in Genesis, 2009), 82–86.

Chapter 8. Legacy

1. The epigraph is from *The World's Most Famous Court Trial: State of Tennessee v. John Thomas Scopes* (New York: Da Capo Press, 1971 [reprint of 1925 edition of trial transcript]), 316, 317.

2. Frank S. Ravitch, *Marketing Intelligent Design: Law and the Creationist Agenda* (New York: Cambridge University Press, 2011), 133.

3. Noah Feldman, *Divided by God: America's Church-State Problem and What We Should Do About It* (New York: Farrar, Straus and Giroux, 2005), 146.

4. Mark Silk, *Unsecular Media: Making News of Religion in America* (Urbana: University of Illinois Press, 1995), 66.

5. On myth in American culture, including the rebel myth, see James Oliver Robinson, *American Myth, American Reality* (New York: Hill and Wang, 1980).

6. Numbers argues convincingly that the model for modern creationist theology was George McCready Price for several reasons: minimal scientific credentials; bragging of not being part of the scientific "establishment"; anti-intellectualism; conservative political agenda; enamored of science. See Ronald L. Numbers, *The Creationists: From Scientific Creationism to Intelligent Design*, expanded edition (Cambridge, Mass.: Harvard University Press, 2006), 81, 101–102, 110, 144–145, 151.

7. George M. Marsden, *Fundamentalism and American Culture: The Shaping of Twentieth-Century Evangelicism: 1870–1925* (New York: Oxford University Press, 1980), 212–215.

8. Ravitch, *Marketing Intelligent Design*, 150.

9. Michael Kazin, *A Godly Hero: The Life of William Jennings Bryan* (New York: Knopf, 2006), 191–192.

10. Cited in Michael Lienesch, *In the Beginning: Fundamentalism, the Scopes Trial, and Making of the Antievolution Movement* (Chapel Hill: The University of North Carolina Press, 2007), 94.

11. David N. Livingstone, *Darwin's Forgotten Defenders: The Encounter between Evangelical Theology and Evolutionary Thought* (Grand Rapids, Mich.: W. B. Eardmans, 1987), 145.

12. William Jennings Bryan, *In His Image*, "The Value of the Soul," reprinted edition (Champaign, Ill.: Book Jungle, undated), 110.

13. Ibid., "The Origin of Man," 64, 74.

14. Marsden, *Fundamentalism and American Culture*, 134–135.

15. Susan Jacoby, *Freethinkers: A History of American Secularism* (New York: Henry Holt and Company, 2005), 17.

16. Ibid., 302.

17. Karen Armstrong, *The Battle for God* (New York: Alfred A. Knopf, 2000), 178.

18. Jeffry K. Hadden and Anson Shupe, *Televangelism: Power and Politics on God's Frontier* (New York: Henry Holt and Company, 1988), 59–60.

19. William A. Dembski, ed., *Darwin's Nemesis: Phillip Johnson and the Intelligent Design Movement* (Leicester, England: Inter-Varsity Press, 2006).

20. Ibid., 123–124.

21. Clarence Darrow, *The Story of My Life* (New York: Charles Scribner's Sons, 1960), 404, 407.

22. Armstrong, *Battle for God*, 140–141, 175–179.

23. Darrow, *Story of My Life*, 409.

24. Richard Dawkins, interview with Bill Maher, at www.outube.com/watch?v=qSr77mv982A, accessed November 30, 2011.

25. Richard Dawkins, "Intelligent Aliens," in John Brockman, ed., *Intelligent Thought: Science versus the Intelligent Design Movement* (New York: Random House, 2006), 92.

26. Lienesch, *In the Beginning*, 203–205; Edward J. Larson, *Summer for the Gods: The Scopes Trial and America's Continuing Debate over Science and Religion* (New York: Basic Books, 1997), 235–238.

27. Armstrong, *Battle for God*, 368–370.

28. Edward Humes, *Monkey Girl: Evolution, Education, Religion, and the Battle for America's Soul* (New York: Harper Perennial, 2007), 53–54.

29. Ibid.

30. Feldman, *Divided by God*, 146–148.

31. Robinson, *American Myth*, xv.

32. Ibid., 6.

33. Edward Caudill, *Darwinian Myths: The Legends and Misuses of a Theory* (Knoxville: The University of Tennessee Press, 1997), xiii–xv.

34. Robinson, *American Myth*, 115, 121.

35. Ibid., 281–282, 287.

36. Ibid., 128–129.

37. Ibid., 310.

38. On the construction of parallel cultures, see Randall J. Stephens and Karl W. Giberson, *The Anointed: Evangelical Truth in a Secular Age* (Cambridge, Mass.: The Belknap Press, 2011).

39. Hadden and Shupe, *Televangelism*, 42–47.

40. On the textbooks, see Judith Grabiner and Peter Miller, "Effects of the Scopes Trial," *Science* (September 6, 1974), 836.

41. Scientists on occasion engage the political aspects of these issues. See, for example, *New York Times*, "Groups Call for Scientists to Engage the Body Politic," August 9, 2011, D1; "Seeking a Missing Link, and a Mass Audience," May 19, 2009, A1.

42. Barbara Forrest and Paul R. Gross, *Creation's Trojan Horse: The Wedge of Intelligent Design* (New York: Oxford University Press, 2004), 215.

43. Christopher Baum, "It's Time to Teach the Controversy," *Skeptic* 15:2 (2009), 42–46. Baum in no way advocates creationism, but argues for using the controversy as a teaching tool.

44. Michael Ruse, *The Evolution-Creation Struggle* (Cambridge, Mass.: Harvard University Press, 2005), 287; Armstrong, *Battle for God*, 369.

Index

Weather Shield®
Premium Windows & Doors

Form No. 1177324-5/07

Weather Shield Mfg., Inc.   PO Box 309   Medford, Wisconsin 54451   715-748-2100
www.weathershield.com

McNamara, James, 20, 32, 34
McNamara, John, 20, 32, 34
Memoirs, 33, 36, 37
Menace of Darwinism, The, 18, 41
Mencken, Henry Louis: *Alleged*, depicted in, 148–149; Bryan, dislike of, 31, 44, 45, 157; Darrow, friendship with, 19–20; fundamentalists, ridicule of, 52, 161, 162; *Inherit the Wind*, depicted in, 26–27, 35; at Scopes trial, 23, 69, 71; on the South, 54, 80, 104
Mere Creation: Science, Faith & Intelligent Design, 92–94
Metcalf, Richard, 21, 22
Miller, Kenneth, 83, 125
modernism: Bryan's, William Jennings, declares it anti-Christian, 36, 38; Bryan, and narrowing of, 18, 156–157, 160–161; Bryan ties it to anti-democracy, 163; creationists, contemporary, fight with, 11, 115; fundamentalist fight with, 1–2, 26, 50, 56, 158; as intellectual movement, 16; Johnson, Phillip, and, 78; in literature, 25; in the press, 23; in Scopes trial, 45, 153; threat of, in nineteenth century, 15; *Wedge Document*, and, 92, 96
Moeller v. Schrenko, 86
Monkey Girl: Evolution, Education, Religion, and the Battle for America's Soul, 4
Moody, Dwight L., 23–24
Mooney, Chris, 3, 110–111, 121
Moral Majority, 29, 66, 67, 72, 73
Morris, Connie, 126
Morris, Henry, 60–61, 64, 66, 129
Morris III, Henry, 129
Museum of Earth History, 104
Museum of Natural History, 137–138
Mystery of Life's Origins, The, 70–74
myth: in American culture, 1, 6, 57, 103, 162; of anti-intellectualism and anti-elitism, 8; in Bible and religion, 41, 71, 90, 160; Bryan, William Jennings, use of, 6, 8, 13, 153, 165; of common-man/Horatio-Alger myth, 164; creationist, use of, 11–13, 141–142, 152, 163–165, 167; Darrow, Clarence, use of, 8, 13, 153; defined, 11, 162; and egalitarianism, 12; in evangelical tradition, 165–166; of frontier, 10, 11–12, 163; of garden, 11, 15, 27; individualism and; 11, 25, 153, 164; Johnson, Phillip, fitted to mythic tradition, 78; of rebel, 138, 153; of science as progress, 11–12, 163; and Scopes trial. 26, 29, 45, 147, 161

Nation, The, 21, 23, 24, 25
National Academy of Sciences, 74, 98
National Association of Biology Teachers, 65–66
National Association of Evangelicals, 118
National Center for Science Education, 81, 99, 129
National Human Genome Research Institute, 90
National Review, 83
National Science Foundation, 61
National Science Teachers Association, 98
Natural Theology, 89
Nature, 82
Neal, John R., 45
Need for Eugenic Reform, The, 150
Nelkin, Dorothy, 5, 63
New Geology, The, 54–55
New Religious Right, 73
New Republic, The, 23, 24, 25, 26
news values, 8–9, 106, 108, 110–111, 166
Newsweek, 60, 63, 95, 109–110, 136
New York Herald, 111
New York Times: on Answers in Genesis, 134; on Ark Encounter, 138–140; Behe, Michael, oped piece by, 110; on Creation Museum, 135, 137; on creationist candidates and politics, 119–121, 123; on creationists, 82; on Discovery Institute, 130; on *Edwards v. Aguillard*, 69; on *Epperson v. Arkansas*, 63; on *Kitzmiller v. Dover*, 97, 102, 107; on *Lemon v. Kurtzman*, 65; on *Of Pandas and People*, 76; on Scopes trial, 48; on the South, 25
New York Times Magazine, 137
New York Tribune, 24
New York World, 22, 52, 111
Nietzsche, Frederick, 38
Nisbet, Matthew, 110–111
Nixon, Richard, 58
NOVA, 83, 125
Novak, Robert, 105–106
Numbers, Ronald L., 2–3, 5, 61

Obama, Barack, 120
Ohio Board of Education, 136
Ohio State University, 136
Olson, Randy, 125–127
Omaha World-Herald, 22, 33
One Million B.C., 137
On the Origin of Species, 2, 15, 92, 137, 150

EDWARD CAUDILL is a professor of journalism and electronic media at the University of Tennessee, Knoxville and the author of *Darwinian Myths: The Legends and Misuses of a Theory.*